HAMILTON EAST PUBLIC LIBRARY
5 0 5 0 8 2 0 1 8 2 4 5 6 3
W9-BZT-708

# THE ASCENT OF INFORMATION

ALSO BY CALEB SCHARF

*The Zoomable Universe:*
*An Epic Tour Through Cosmic Scale,*
*from Almost Everything to Nearly Nothing*

*The Copernicus Complex:*
*Our Cosmic Significance in a Universe*
*of Planets and Probabilities*

*Gravity's Engines:*
*How Bubble-Blowing Black Holes Rule Galaxies,*
*Stars, and Life in the Cosmos*

# THE ASCENT OF INFORMATION

## BOOKS, BITS, GENES, MACHINES, AND LIFE'S UNENDING ALGORITHM

Caleb Scharf



RIVERHEAD BOOKS  NEW YORK  2021

Riverhead Books
An imprint of Penguin Random House LLC
penguinrandomhouse.com

LIBRARY OF CONGRESS CATALOGING-IN-PUBLICATION DATA
Names: Scharf, Caleb A., 1968– author.
Title: The ascent of information: books, bits, genes, machines, and life's unending algorithm / Caleb Scharf.
Description: New York: Riverhead Books, 2021. |
Includes bibliographical references and index.
Identifiers: LCCN 2020046554 (print) | LCCN 2020046555 (ebook) |
ISBN 9780593087244 (hardcover) | ISBN 9780593087268 (ebook)
Subjects: LCSH: Information theory. | Human-computer interaction. |
Human-machine systems.
Classification: LCC Q360 .S293 2021 (print) | LCC Q360 (ebook) |
DDC 003/.54—dc23
LC record available at https://lccn.loc.gov/2020046554
LC ebook record available at https://lccn.loc.gov/2020046555

Printed in the United States of America
1st Printing

Book design by Daniel Lagin

*To the 130 million books that came before this one*

# Contents

# THE ASCENT OF INFORMATION

# 1

# OUR ETERNAL DATA

*Nature produces those things which, being continually moved by a certain principle contained in themselves, arrive at a certain end.*

—Aristotle, *Physics*, Book II, 350 BC

In this instant, a precious one-second span out of the four and a half billion years Earth has existed as a bejeweled sphere of complexity and dynamism, I am gripped by one puzzle only: Can those really be tears glistening in the eyes of the museum guide standing in front of me?

Perhaps the guide, too, is momentarily caught up in thoughts of the rich tapestry of existence, brought to an emotional precipice. Or, since this seems improbable, maybe some part of her anatomy is being chafed as she resolutely delivers what must be an extremely well-worn line ". . . and this . . . is where the young William would have slept."

Before I can adjust my gaze to follow, I'm diverted by the sound of hysterical giggling coming through the open window from the street, where a dozen tourist groups cluster. Unfazed, our dedicated mentor on all matters of the youthful Shakespeare presses on to deliver a final heart-stopping temptation: "You can read more about everyday life in this house on the informational placards."

Sure enough, pieces of laminated text are dispersed across the room, strategically located where you might try imagining a domestic scene from four centuries earlier. Laser-printed, I think to myself. Nice fonts.

Here in Stratford-upon-Avon in the UK, the theme is all Shakespeare, all the time. I've dutifully parked our car outside the city center and forced my family to trundle inward on the Park-and-Ride bus, past all the usual trappings of twenty-first-century life in the West. There's a hairdresser's, there's a pub, there's a hotel, an Indian restaurant, a chain sandwich shop. And there is poster after poster for this season's performances at the Royal Shakespeare Company.

Having descended through these layers of contemporary civilization, we are now well and truly prisoners for the day. First up on our hastily conceived activity list has been the house that was the birthplace in 1564 of William Shakespeare, son of John and Mary. Now, with the placards politely read, we're back to walking around the streets, squeezing by yet more tourist groups of every conceivable nationality. I hear at least half a dozen languages, and remind myself that Shakespeare's works must have been translated into all of these tongues and many more.

Along the admittedly quaint streets are boundless opportunities to accumulate staggering numbers of Shakespeare-related knickknacks. Busts of the bard in all sizes and color schemes, haphazardly molded or carved somewhere in China or Indonesia. Postcards, trinkets, banners, and T-shirts with slogans: "All the World's a Stage," "Will Power," "I would challenge you to a battle of wits, but I see you are unarmed." Then there are the Hathaway Tea Rooms and Bakery, the Creaky Cauldron, The Pen and Parchment, and many more establishments reminding us not-so-subtly of where we are, and that we might need to be refreshed or further amused.

Finally, after wading through another batch of Elizabethan houses, and more points of sometimes questionable historical interest, we make it to the place I really wanted to see most of all: Holy Trinity Church, and

Shakespeare's grave. Not because I'm harboring any particularly ghoulish dreams, but because I want to see Will's epitaph in the flesh—so to speak.

And it's a great epitaph. Carved into the flat flagstone inside the church, right up at the foot of the altar, are the words:

*GOOD FREND FOR IESVS SAKE FORBEARE*
*TO DIGG THE DVST ENCLOASED HEARE*
*BLESTe BE Ye MAN Yt SPARES THES STONES*
*AND CVRST BE HE Yt MOVES MY BONES*

The orthographic conventions used are a little tricky for modern English-speaking brains, so here's a translation:

*Good friend for Jesus sake forbeare,*
*To dig the dust enclosed here.*
*Blessed be the man that spares these stones,*
*And cursed be he that moves my bones.*

Regardless of debates about whether or not Shakespeare himself penned these exact words, this has to be one of the most memorable and original passages to ever adorn a gravestone. Its ambiguous tone—playful, but also utterly menacing—keeps your attention. And it's easy to see it as a final nose-tweaking taunt of authority—a knowing reminder to the church that it had better not repurpose this spot, or else. Or else, I think, you'd not have the endless streams of eager tourists and coins in the donation box.

Standing there, feeling slightly soiled from the day, I am struck by the sheer absurdity of it all. This one human, William Shakespeare, wrote a bunch of stuff some four hundred years ago, and that stuff has radiated outward, in space and time, like a brilliant pulse of light spreading into the cosmos. His words have been reproduced and copied on an astronomical scale. Those words have prompted new words—writings of critics, of fans, of historians, of schoolroom essayists, and of me as I think these thoughts and craft the phrases written here. Although the bard himself was definitely not immortal, a fact his grave clearly testifies to, his ideas and creations might be. Yet the conceptions and stories that William Shakespeare converted into written matter never existed inside him as anything more than synaptic connections and electrochemical pulses. This information was not encoded in his DNA. He could not biologically bequeath it to his descendants. There were no heritable genes for his thirty-seven plays.

All of that information is still here, though. Outliving him, and influencing our lives all this distance down the human timeline. As I flex my tired feet I am acutely aware that on this ordinary day in the twenty-first century my actions can be directly attributed to Shakespeare's informational remains—his data from hundreds of years ago.

Shakespeare didn't reach across the ages to me personally. The data he created—his plays and sonnets and his epitaph—is what's affecting me and my family, and the two to three million people visiting Stratford each year, as well as billions of humans now and in generations past. And it's no longer just his first crop of data that's influencing the world. His original works are well preserved, but now there is a vast ocean of descendant material. All those scholarly analyses and all those Hollywood reimaginings packaged so as to obscure their Shakespearean sources, from *The Lion King* (*Hamlet*) to *10 Things I Hate About You* (*The Taming of the Shrew*). And again and again, his turns of phrase are repeated and redeployed. We still talk about shuffling off this mortal coil, being pure as the driven snow, and breaking the ice, all with a heart of gold.

Some aspects of this phenomenon are captured by the famous (or possibly infamous) concept of *memes*, a label invented by the ethologist and evolutionary biologist Richard Dawkins in his hugely influential 1976 book *The Selfish Gene*. Although the basic idea had been around before in different guises, Dawkins' neologism "meme" was, without any irony, the term that propagated itself in the popular consciousness. A meme is a unit of cultural transmission, or of imitation and replication. As originally envisioned it is a dynamic "replicator," constantly being propagated by and stored in living minds. I'm going to come back to this later in much more detail, but for now I'll simply point out that while the spread of Shakespeare's ideas and phrases is decidedly meme-like, it arguably stretches beyond that. His data not only seems to have a life of its own, it has manifested physical structures of its own. Much like his carved epitaph, it persists regardless of whether or not there are human vehicles to carry it at any given moment.

Of course, Shakespeare is only one example of how this happens. Our entire species is drenched in data. This is one of the most peculiar and possibly unique features of humans: we carry vast amounts of information externally to our biological forms. We've been doing this for a long time, and we are very good at propagating that information into the future and making use of it. And just as with Shakespeare himself, today our carried data far exceeds the root information contained in all of our DNA. There is so much data that it spills across the world: in books and electronic media, as well in all of the artifacts that go hand-in-hand with those data repositories, from brick-and-mortar libraries to fiber-optic networks.

The philosophers Andy Clark and David Chalmers have gone further to observe that aspects of our physical environment, as well as our languages, might function as an "extended mind." Those hastily scribbled notes or architectural models or physics experiments could represent parts of an extended cognitive system. Perhaps our cognition is really the sum total of what happens inside our brains and what happens outside.

These observations, some that at first appear quite simple, mark the beginning of the story that I will tell in this book. This story turns out to be vastly more complex and surprising than could ever have occurred to me as I stood a little woozily in the nave of Holy Trinity Church. It's a tale that first leaps from Shakespeare to the energetic burdens of maintaining an information-rich world. We can understand that burden through nothing less than the fundamental physical, mathematical properties of matter and a special measure of order and disorder in the cosmos. From the earliest stirrings of human communication and informational invention we'll travel all the way to the grandest possible perspective: that of life in its cosmic setting, encoded in atoms and molecules, but pressing up against even deeper substrates of reality. In between these extremes are a series of many interlocking puzzles. Puzzles that include the nature of meaning in information and its relationship to biology, and what that meaning costs our planet, as well as the emergence of a machine world . . . and, ultimately, what it is about the workings of the universe that actually compels us to write, build, and compute in the first place.

I'm going to explore how the forces of evolution and the mechanisms of natural selection help connect all of these waypoints. We'll probe into the fundamental nature of information itself—as something of substance rather than abstraction, a part of the laws of nature. Navigating this trail will challenge us to think about what we even mean when we talk about the nature of life and living things—and whether our conceptions of ourselves and our intelligence are due for a drastic, and disturbing, overhaul.

The central phenomenon in this journey needs a name. I've come to call it the "dataome," similar to the concept of a genome. Dataome is a combination of the Latin-rooted "data" with "ome," which has its origins in ancient Greek and means a mass of something, or the complete class of substances for a species or an individual. Other names might include "exodata" for data or information existing "outside," or "datasphere" as a parallel to the biosphere (and distinct from the "technosphere," which is the physical environment made or modified by humans). The dataome is all of the non-genetic data we carry externally and internally. It encompasses William Shakespeare's works and their progeny, as well as everything else that we know of human information. It is, as I'll argue, far from being merely derivative in nature, and far from being a passive component of Earth's cascading systems of life, energy, and material structure.

Although we often use the words "data" and "information" interchangeably, strictly speaking they do carry different meanings. In computer science data is typically considered to be raw material, facts, and figures. In that sense data corresponds to pieces of information. But real *information* is that data organized and assembled, and structured to provide meaning and context. A set of data might be the list of words "all," "the," "world's," "a," and "stage," but the information of that dataset is "All the world's a stage."

Similarly, the dataome comprises data with different fundamental characteristics and different degrees of information. For instance, there is data that, while not genetically encoded, is only ever held in biological structures. The majority of your personal memories would be in this category, never straying beyond the confines of the dendrites and synapses of your brain and vanishing when you die. Then there is data that might persist as memes, perhaps never committed to physical form outside of human brains but shared and maintained regardless of how many generations come and go. And there is data that is fully encoded in the physical world external to human bodies and seemingly unconcerned with whether or not it is interacted with, be it in letters, books, films, music, hard drives, or silicon chips. But there is also information in these structures themselves. The arrangement of a book is a manifestation of an idea, a design made of data regardless of the words on its carefully measured pages. A library is a further elaboration of that idea, a brick-and-mortar object originally encoded in an architect's data that now sustains and curates other data and helps create information from that data.

All of these things seem like they must originally be set in motion by our genetic material, the expression of those genes, our genotype. In biology we refer to the phenotype, the observable characteristics of a living thing due to the interaction of its genotype with the environment. A phenotype encompasses an anatomical form and its appearance. We also sometimes talk about the *extended* phenotype, another concept advanced by Richard Dawkins, back in 1982.

The extended phenotype is the sum of all the changes that an organism makes to its environment: the structures it builds, like burrows and dams, all the way to cathedrals and cities. Even the parasitic impacts on others can be included, and there are astonishing examples of that kind of extended phenotype. A certain species of nematode (a tiny roundworm) infects ants and distends their bodies to look like ripe red fruits. Berry-loving birds that usually avoid ants will then eat them, and disperse the nematode eggs in their droppings, which more ants harvest for their larvae, reinfecting a new generation. The fruit phenotype of the ants is actually an extended phenotype of the nematodes.

Other ideas that try to capture the extended properties of genes expressing themselves in the world include the concept of "niche construction." This refers to the way living things can manipulate their environment and pass that manipulated system on to subsequent generations—just as we might build a house with the intention for our children and grandchildren to benefit from it.

But, as we'll see, the dataome isn't fully describable with concepts like memes, extended phenotypes, or niche construction. These phenomena are definitely related, even a significant part of what the dataome is, but I'm going to argue that the dataome is much more. One simple clue to this is the very unfiltered way in which our human dataome can interact with our entire species, not just the individuals who contribute to it.

Beyond anything else, the concept of the dataome is a new lens for looking at the world. It remains to be seen whether, as I've conceived it, the dataome offers us genuinely predictive, testable explanatory powers—as a good scientific hypothesis should—but it very definitely raises some of the most challenging and compelling questions about our nature, and the nature of life in general, that we have yet faced. By the end of this book I hope to have some answers.

Certain critical elements of this story have long been topics of debate in some specialized circles. A number of thinkers over the years have asked whether information itself may be the fundamental currency of the universe—superseding our comparatively parochial ideas of what makes biology and chemistry, or even physics, tick.

Notably, the physicist John Archibald Wheeler (who helped coin the term "black hole," among many other more substantial accomplishments) explored the notion that the ultimate nature of physical reality is inextricably linked to observation and experimental interrogation. The basis for this lies in quantum physics, and the bizarre laws of quantum mechanics. In the most successful framework we have for describing quantum phenomena, the world is made of fields (akin to electrical or magnetic fields) whose energetic oscillations or ripples can manifest subatomic particles. But prior to being observed or interacted with, these particles exist in a blurry superposition of possibilities in space, time, and state. The very act of observation or interaction is what causes their properties to snap into focus. In other words, this is a *participatory* universe of yes/no information, in which, as Wheeler put it, we get "it from bit."

Wheeler used a clever illustration, a group game of twenty questions that he called "negative twenty questions." In this variant of the usual play, the guesser asking yes/no questions believes the group being interrogated has a single item in mind that they've all agreed on in the guesser's absence. In actuality, each person can start with whatever they want in mind, and the ultimate answer is gradually shaped by the questions. Each response anyone gives must be consistent with all previous responses. So if the first question is "Are you thinking of a human?" and the answer is "no," then all subsequent answers have to be compatible with that. (Anyone who had a human in mind must change that.) If the next question is "Is it an insect?" and the answer is "yes," the options begin to narrow. The extraordinary thing is that if you try playing this game you will eventually converge onto a specific answer—an item that uniquely satisfies all previous answers. In a very real way, that item has snapped into focus not because it already existed, but because the questions were participatory: they *made* it come into existence. It came to be from yes/no, from 1 or 0.

If we take this view of reality, then decisions of yes or no, red or blue, don't just enable analysis of the world. At the deepest level they actually create phenomena and their functions. If you think that's head-spinning, well, yes, it is. But the overarching idea that information is something real, something that actually pulls the puppet strings on the world, and on us, is actually not so hard to grasp.

That is precisely the nexus where I think that the concept of the dataome—the sum total of our internal and external information—has the most to tell us. Because at its core, the dataome is not just about what we get from our data, it is about what our data gets from us.

To really understand the dataome we also have to deal with challenges that parallel those faced by evolutionary biologists. Life on Earth today is the result of four billion years of evolution—a restless and energetic party of experimentation and interaction, with untold quadrillions of individual organisms and their complex molecular innards as guests. Just as it can be hard to decipher the previous night's activities when faced with the morning after's detritus and a blurred memory, when it comes to the precise history of life on Earth we have to study it through a billion-year hangover. Reconstructing the actual historical sequences of events and branching possibilities is tough when most of what we see are the survivors, the pruned remains of more species that once were than are here now.

As a result, we often have to resort to delicate extrapolation to infer what possible pathways and strategies life has explored and suffered through. The dataome can require similar approaches. Human information today is the product of at least a couple of hundred thousand years of accumulation and evolution. But there is also a critical difference from biological history: the dataome more explicitly records its own development. That may let us accurately reconstruct its relationship to us, and learn what it really is.

## Hello, World

When did the human dataome come into existence? At some point members of our ancestral hominin community, including *Homo sapiens*, *Homo neanderthalensis*, *Homo erectus*, and *Homo denisova*, began to propagate sets of information across the spaces of Earth and through time. But there was something about these repositories that distinguished them from what had come before, and from what other species do to organize and propagate the information that is unique to them.

Some bird and insect species build signature nests or hives. But we don't really know whether a bird builds nests because of instinctive tendencies or because it learns, or through a combination of both behaviors. Recent research on a subspecies of weaver bird, renowned for making intricate nests by weaving grasses into highly complex three-dimensional structures, suggests that some of the birds may learn to do this rather than being born with the instinct to do so. But that learning may not exclusively involve their observing other nests or other birds. It may be that they figure it out iteratively—improving each time they build a nest for themselves, with an instinct for judging their success or failure. There doesn't seem to be strong evidence that weavers get advice from each other, or even that they observe other nests and think, "I like that, I'll copy it." The avian dataome, if it exists, may be quite limited in scope.

Hominins have evidently done things differently, and the evolution of complex language must have been central to this. Exchanged observations, ideas, skills, and stories seem likely beginnings for the human dataome: information held in biological brains, not necessarily recorded in any other medium, but propagated into the future, sustained and augmented over time. In this proto-dataome the phenomenon of memes represents a particularly dynamic subset of infectious, self-propagating information.

Modern humans are also in many respects like a global monoculture, and that could be important for a dataome. Genetic studies made over the past ten years suggest that if you pick any two humans at random, from anywhere on the planet, their DNA will differ by no more than about 0.1 percent to 0.6 percent. At the very most, 6 in every 1,000 nucleotides will be different, implying a person-to-person variation in at most some 20 million nucleotides in our 3-*billion*-nucleotide-long genetic code. These nucleotides occur as four kinds of molecules (adenine, guanine, cytosine, and thymine) that make the "rungs" in the twisted double-helix ladder of DNA, the thread of life on Earth. They are literally the letters in biology's alphabet, nestled among the roots of how life encodes information, form, and function.

For comparison we can look at our closest evolutionary relatives: the chimpanzees. Common chimpanzees in equatorial Africa fall into one of four recognized groups: western, central, eastern, and Cameroonian. Even though these are all geographically close, researchers discovered in 2012 that the average genetic difference between any of these populations is about 1.2 percent. That's significantly larger than the maximum human variation across the entire planet. Separation by the Congo River seems to result in greater differences between chimps than between humans from entirely different continents.

Results like these, together with the study of human mitochondrial DNA (the very seldom–varying, maternally inherited genetic code that helps our cells transform energy), have helped lead us to a picture of human ancestry that is markedly distinct from that of our closest relatives outside our species. Two key characteristics of modern humans are that we migrated back and forth across the world, and that we seem to have experienced population bottlenecks. Both of these things have restricted our genetic variation.

Perhaps most significant is how episodes of migration, like those across the Bering Strait to the Americas, may have pruned our ancestors down—leaving smaller, more genetically related groups to populate entire continents. This so-called "founder effect" helps explain why today there is actually more human genetic diversity within sub-Saharan Africa than between populations anywhere else in the world.

This backdrop of restricted genetic diversity across the history of *Homo sapiens* seems like it could be important for the development of a dataome. Put simply, compared with other species our remarkable genetic uniformity might have reduced differences in learning and behavior and thereby affected the ways in which we've generated, curated, and propagated information. This suggests that the extraordinary diversity of human data across time is not a consequence of fundamentally different minds. Nor is it that we've necessarily gotten smarter over time. Rather, it is that our systems of information and teaching have been built on a consistent biological bedrock. That steadiness has enabled our dataome to persist and evolve while undergoing constant experimental tweaks and refinement for thousands and thousands of years.

But tracking down when external data started to really become a big piece of human existence is not easy. This is especially true for data that was never committed to a physical form, such as the information held in oral traditions. Where we do find that data, it has remarkable stability and persistence. Today the most famous examples of oral storytelling come from the Aboriginal peoples of Australia. Having settled across the continent from upwards of 50,000 years ago, many groups developed sophisticated techniques for maintaining collective knowledge. These techniques have shown themselves to be highly accurate—storing error-free data for an astonishing number of generations.

In southern Australia the Tjapwurung people maintained a story about a specific hunt that took place for a species of giant bird that no longer exists. This was probably *Genyornis newtoni,* a fowl that stood as much as two meters tall and lived until perhaps 30,000 to 10,000 years ago, with a lineage going back at least eight million years. Not only did the Tjapwurung story detail the perils of tackling these large, strong birds, it had enough specifics about the hills and rocks of the landscape and the conditions of locally erupting volcanoes to enable modern geological measurements to be used to pin down and confirm the tale's historical timeline.

Another example involves stories tracking a sixty-meter sea level rise witnessed by dozens of generations across the shorelines of Australia some 10,000 years ago. We know that this change indeed took place; it was the Holocene sea level rise between about 12,000 and 7,000 years ago, due to the rapid deglaciation and ice-shelf melt at the tail end of a warming period following the Last Glacial Maximum. Remarkably, multiple lines of the oral histories passed down by indigenous Australians, across some 400 generations, have recorded details of ancient shorelines and now-submerged islands with such fidelity that these places can be precisely identified today.

The secret to the maintenance of these accurate accounts is a brilliant system of cross-generational cross-checking. If an adult is taught a story by their aging parent, then their children and their nephews and nieces are given the parallel task of fact-checking and keeping the recipient's memory accurate. As a result, there is kin-based cultural responsibility and there are at least three generations in possession of the stories at any given time. In other words, more than 10,000 years ago humans had developed a highly effective error-correction system for their precious data.

By comparison, our early efforts to capture information in ways that are truly external to our biological forms can actually seem a little more haphazard. Take, for example, one of the earliest recognized full writing systems: Sumerian cuneiform.

The Sumerian language itself is interesting. The Sumer civilization is the oldest one we know of in the region of southern Mesopotamia (now southern Iraq). There is evidence of its roots back in the prehistoric period around 5,000 BC. What many of us are taught to think of as the transition from prehistory to history is connected to the explicitly informational records of that civilization, produced sometime around 3,200 BC.

But Sumerian is special; it's known as an isolate—it has no obvious relation to past, present, or coexisting languages. It was genuinely unique and it had a structure that is called *agglutinative*.

In essence this means that the language is *synthetic*. The units of meaning (termed *morphemes*) are put together to make words. And those units don't get changed; they're kind of like fundamental building blocks of meaning. So single words can become increasingly long as their meaning becomes more complex. Turkish is an agglutinative language, where a word like "yaramazlastIrIlamIyabilenlerdenmissiniz" is a combination of pieces like "yaramaz" (not useful or naughty) and "las-tIr-Il" (acting like or being like) strung together to tell us that "*You seem to be one of those who is incapable of being naughty.*" By contrast, a language like English is mostly *analytic* in nature: meaning can hinge on word order in a sentence, and there are specific words for prepositions and participles.

For Sumerian, being agglutinative was intimately connected to its recorded form. The cuneiform method of writing involved repeatedly pressing similar triangle-like shapes into a soft material—cuneiform literally meaning "wedge-shaped." So sticking together chains of symbols, agglutinating them, works very naturally. Two million or so cuneiform clay tablets have been unearthed by modern excavations, charting both the history of Mesopotamian civilizations and the evolution of the written language itself. Cuneiform wasn't the original form of recording, which was pictographic. The precise triangles and lines of cuneiform represents an evolved version, an abstract streamlining of what were once cartoonish and variable pictures. This must have increased efficiency and presumably overall readability, helping reduce errors and increasing the utility of the information.

Cuneiform persisted for 3,000 years as a robust form of data recording. Versions of the script were also adapted and used over time for many other languages. In addition to Sumerian itself, cuneiform ended up being used for Akkadian, Babylonian, Assyrian, Elamite, Hittite, Urartian, and Old Persian. For these groups, cuneiform held the history of business

The evolution of Mesopotamian writing, from pictogram to abstract, for the word *head*.

relationships, accounting, recipes, laws and legal agreements or disagreements, and notes of medicines and science—pretty much anything you can imagine a sophisticated population putting on record.

Despite that robustness, cuneiform appears to have been utterly supplanted by social changes. It effectively went extinct around 300 AD, when the neighboring Sassanian Empire, known to be hostile to foreign religions, seized Babylon and shut down the temples where scribes had been trained. That act probably stalled certain scientific and intellectual development by centuries, or at the very least thwarted the trajectory they had been on. Yet the Sassanian period also represented a peak of the ancient Iranian civilization and creative culture during that era. These invaders were the polar opposite of ignorant; they were bent on having things their way—including the way information was stored and probably, most critically, who was doing the storing and retrieving. There was already power in data.

As a result of this regime change, collections and buried stashes of cuneiform tablets sat untranslated, and untranslatable, from about 300 AD until the nineteenth century. There was no one left who knew how to read them. But starting with the curiosity of a few individuals in Europe in the 1700s, and accelerating in the period of the 1800s, cuneiform tablets eventually became sought-after collectibles. Their allure was precisely due to their mystery. Their unknown content was irresistible.

Numerous scholars worked on the problem, gradually piecing the script together. A particularly important moment in the resuscitation of the ancient cuneiform dataome came in 1872 when the self-taught British Assyriologist George Smith (he'd left school at fourteen) managed to translate a crusty scrap of a tablet. Here he found a familiar-sounding tale of a Great Flood. But it wasn't the one taught in every Western schoolhouse. It was a Babylonian tale that included multiple gods, a Noah-like character called Utu-napishtim who made a boat, and a final receding of the floodwaters scouted out by birds. It was a piece of the Epic of the Sumerian king Gilgamesh, who ruled Uruk sometime between 2800 BC and 2500 BC.

Not only did Smith's efforts fully ignite the world's curiosity about the mysteries of cuneiform, they also helped usher in a dramatic shift in historical perspective, making it clear (to the more open-minded) that the Biblical view of the past was neither the oldest nor the last word. Instead, it was founded in the myths of much earlier civilizations.

The story of cuneiform is fascinating, but if we look at all of this through the lens of the dataome it gets even more intriguing. First is the observation that data need not be constantly engaged with. It can sit, in inert tablet form, for thousands of years. Yet once it re-enters the current human culture it takes on new value, yielding insights and knowledge that were almost certainly never anticipated. I'll come back to this critical aspect of the dataome later on.

Second, it was the very inaccessibility of this information—untranslatable, mysterious—that intrigued people thousands of years later. It is a great example of the allure of information for humans, the sense that there could be something exciting or useful in what are otherwise simply markings on pieces of dry clay.

Finally, this story can be interpreted in an even more provocative way. It could be seen as evidence that the dataome directly influences human behavior and drives our expenditure of time and energy. In the case of cuneiform, part of the dataome eventually found a way to propagate itself off of clay tablets and onto paper and hard drives.

## All the Pretty Things

There are so many, sometimes surprising ways to preserve information. In Andean South America a number of human groups and societies once used devices called "khipu" or "talking knots" ("quipu" in Spanish). Their precise origins in this region are hard to pin down, but the Incas were definitely using khipu extensively from around 1400 AD, and some archaeological remains point to the existence of khipu five thousand years ago.

A typical khipu consists of a long textile cord with other cords (sometimes called pendant cords) attached to it, and with knots tied along their own extents. These cords can be different in lengths, materials, colors, and sometimes have further cords attached to them in turn. The knots themselves come in some variety. There are "S" or "Z" knots that have the same underlying structure, but with different orientations (like a left and right knot). There are figure eights with a twist and there are "E" knots, as well as long knots in which the cord wraps multiple times around itself, and simple single knots.

Many, perhaps most, khipu appear to be sophisticated data storage for numbers. Khipu have been found in long-lost settlement sites where commodities such as peanuts, chili peppers, or corn were assessed and stored. The knot structures form a basis for counting these goods. In one interpretation, or reconstruction, by the researchers Marcia Ascher and Robert Ascher, the positioning, clustering, and type of knots correspond to specific mathematical meanings. For example, powers of ten are indicated by location down the hanging cords. The number "30" might be recorded as three simple knots in a row in the "tens" section of the cord. The figure-eight knot seems to represent a "1" and also delineates separate numbers on a single hanging cord.

It also appears likely that a khipu had an abacus-like function (or perhaps it would be fairer to say that the abacus has a khipu-like func-

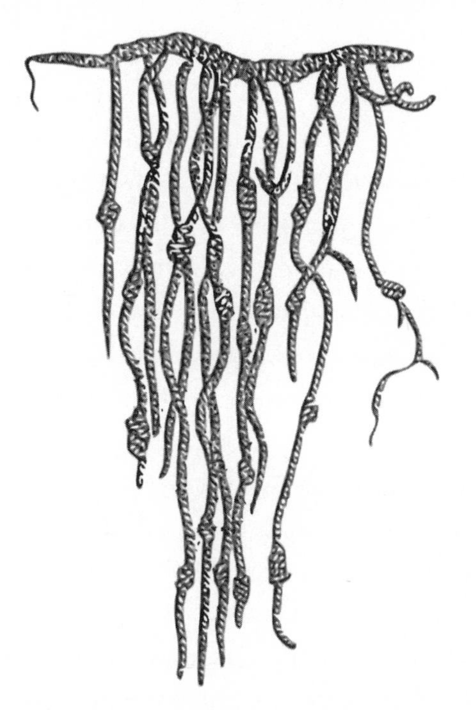

Drawing of a khipu from 1888, in fourth edition of the Meyers *Konversations-Lexikon*.

tion). Sometimes a cord will indicate the sum total of the next several cords, and sometimes the sum of sums. This could clearly also be a method of error control, not unlike the checksums that we use in computer programming to verify that a code has not been altered or affected by mistakes.

But khipu may have had a whole other application as well. A number of researchers have suggested that khipu could hold significantly more sophisticated information. Cord colors, and particular knot structures and clusters, might have denoted the location where the khipu was made, or even the time at which it was knotted. Perhaps the khipu keepers had the equivalent of a signature too—a useful way to correlate data in the future. Other suggestions include the idea that khipu were used to store musical or theatrical performances.

There are also indications that as the indigenous cultures of South America faced the invasive populations of Europe, particularly the Spanish from the late 1400s onward, khipu became tools of resistance as well as of religious conversion. In one particularly delightful story, recorded in the 1600s by a priest attempting to bring Catholicism to a southern Peruvian village, the khipu are described as "tangles for their souls." It appears that the locals used khipu to track and make confessions for their supposed sins. But they would happily borrow and falsify the khipu, making life enormously (and perhaps satisfyingly) difficult for the priests accepting their confessions.

Today, while there don't seem to be any people alive who can read or use the ancient versions of khipu, these knotted cords are still a part of social cohesion. There are communities in Peru that incorporate preserved khipu into important ceremonies, such as when village officers take up their jobs, or leave them. Khipu are brought out for display and to be draped on the lucky officials, not unlike the many ceremonial artifacts used elsewhere in the world to add some pomp and seriousness to the proceedings.

But from the perspective of the dataome perhaps the most important point is hidden in plain sight. Just as with cuneiform and its clay tablets, our modern fascination with the khipu and their data has enabled a migration onto modern storage devices. Online at Harvard University in the US you can visit the Khipu Database Project, where khipu images and databases are freely available. As I write this, elsewhere on my computer screen is a spreadsheet displaying the information from one 200-cord khipu, noting cord lengths and properties, the knot clusters and their interpreted numerical values. There are more than 500 other khipu datasets here, a digitized repository of khipu whose full meaning we still do not know.

To me this is a remarkable example of the persistence of human information, and of the seeming capacity of that information to reinvent itself and to re-infect (or perhaps re-inoculate) the world. Today a piece of textile with a set of knotted cords that was assembled by a merchant in Peru centuries ago to track their crops or their confessions can be accessed by any human with an internet connection. But for that to happen the modern world had to happen. Science had to decode the possibilities of electricity and logic gates. Human industry had to learn to fabricate exquisitely finely etched silicon and microprocessors, the modern-day forms of generalist khipu. Societies across the planet had to evolve, to feel compelled to mold and reinvent their environment with energy and raw materials on a scale unforeseeable by previous cultures.

## Right Time, Right Place

Cuneiform and khipu worked well in their time. They were, if you will, successful evolutionary experiments that merely succumbed to changes in their environment and the limitations of their niche roles. Not everyone in ancient Mesopotamia or Peru knew how to write or read with cuneiform or khipu, nor were they even allowed to. Had that been different or had certain technical innovations occurred, their stories might have ended up with an alternate ending. Even the printed page had to break out of its niche in order to become one of the most important evolutionary branches of human data dissemination.

Most popular, Western-centric accounts focus almost exclusively on the goldsmith Johannes Gutenberg as the inventor of the movable type printing press sometime around 1439 in the city of Strasbourg. Later, in 1455, back in his birthplace of Mainz, he began mass printing his "42-line Bible." But it is also well known that different elements of what came together in Gutenberg's printing press had existed much earlier in a variety of different forms.

Before 3000 BC the Mesopotamian culture had made use of cylindrical seals for rolling out images onto clay tablets. In China, by around 800 AD, mechanical ink printing using carved woodblocks was churning out copies of works such as the *Diamond Sutra*, describing an interaction of the Buddha with a follower and a dialogue on the perfection of insight and the nature of reality. An 868 AD copy of this sutra is often described as the oldest surviving complete printed book, which comes with what could be considered a statement of open, public domain copyright printed on its 17-and-a-half-foot-long scroll.

By about 1000 AD in China it's apparent that a form of movable type had been invented. Characters were made out of ceramic or wood that could be rearranged and fixed on an iron plate with a sticky paste to make a page. By the twelfth century, bronze movable type was being used for documents and even money printing, with the technique being refined through experimentation with metals like copper and even tin. Across Asia, in Korea and Japan as well, movable type ink printing was a robust enterprise well before the 1400s.

In Europe, wooden block printing was also well established before Gutenberg. But part of what enabled Gutenberg to evolve the printing process to a new level was his invention (undoubtedly itself developing out of work with his contemporaries) of techniques for rapidly and robustly making the type itself. This included a durable lead, tin, and antimony alloy for the metal characters; an approach to quickly precision-mold new type; and the use of oil-based inks. Or at least this seems to have been the case. In the numerous histories written about Gutenberg and his creations there is some ambiguity over exactly how his approach developed and even over what books and materials he first churned out. The fact that he was a bit of a financial disaster as a businessman (although later feted and supported in light of his accomplishments) seems to have obscured some of the details.

None of which downgrades the incredible impact of the novelty of the Gutenberg press. The ability to rapidly and (comparatively) cheaply mass-produce copies of written material was an astonishing catalyst of social and scientific change. From the Renaissance to the Age of Enlightenment, and on to today's global information civilization, the quickly produced printed page has been a central factor. Knowledge once held tightly as a component of power by religions and rulers was no longer so exclusive. Equally, dissemination of a religious or philosophical worldview became vastly easier.

Movable type may have taken centuries to take hold of our species' dataome, but once it did there was no looking back.

# 2

# THE BURDEN OF AN IDEA

*But if your theory is found to be against the second law of thermodynamics I can give you no hope; there is nothing for it but to collapse in deepest humiliation.*

—Sir Arthur Stanley Eddington, 1927, Gifford Lectures, Edinburgh

If movable type was like gasoline sprayed on the fire of human data, some exceptionally volatile piles of kindling have also existed. Let's look at a few more numbers for our old friend William Shakespeare. At the time of his death he had written a total of thirty-seven plays, a hundred and fifty-four sonnets, five long narrative poems, and likely collaborated on several other pieces associated with different authors. Those thirty-seven plays alone contain a total of 835,997 words. In the centuries that have come after his corporeal life an estimated two to four billion printed copies of his plays and writings have been produced.

That last figure is a pretty rough and conservative approximation. Many of his plays have been translated into more than eighty languages at one time or another, and it's impossible to know how many copies or partial copies have been made over time to hand out to aspiring thespians or leery teenage students of English literature. All of these volumes and versions have been composed of hundreds of billions of sheets of paper acting as a canvas for more than a quadrillion ink-rich letters. Even though today we might switch on a tablet's screen or an e-reader's light and feel momentarily good about saving some trees, words on physical paper still dominate, feeding our global appetite for written material.

The production of printed books or manuscripts has been on a steady climb since Gutenberg's innovations in the 1400s. This is true both in terms of the number of unique titles and the absolute number of copies made. Between 1550 and 1600 an estimated 140 million copies of books were printed around the world (supplying a total human population of around 500 million). That rate of production has grown steadily; currently in the United States alone about 675 million print books are sold each year. If we take that number as a lowest possible limit, since it's just one country and just the number of books sold, this implies that today, as a species, we're making physical books at a rate that's *more than 53 times higher* than two hundred years ago. But the human population has only grown by a factor of 7 or so during that period.

Books are surprisingly demanding little things. In the case of Shakespeare, across time billions of copies of his texts have been physically lifted and transported, dropped and picked up, held by hand, or hoisted onto bookshelves. Each individual motion has involved only a small expenditure of energy, but that has added up over the centuries. When I crunch the numbers, assuming a typical mass of a few hundred grams for a paper copy of a play and its average vertical motion through about a meter, it is possible that altogether the simple act of human arms raising and lowering copies of Shakespeare's writings has expended over 4 trillion joules of energy. That's equivalent to combusting several hundred thousand kilograms of coal.

The usage burden doesn't stop there. Additional energy has been utilized every time a human has read some of the bard's words, firing the electrochemical signals of neurons in our brains and causing neural capillaries to flush with oxygenated blood. That oxygen, incidentally, is generated by photosynthetic organisms spread across Earth, from land plants to microscopic marine diatoms, all capturing stellar energy that has already spent an average of 100,000 years diffusing outward from the Sun's core in the form of photons scattering through dense stellar matter before escaping for their brief trip to Earth. And that stellar energy first arose from the conversion of mass into energy in nuclear fusion reactions that predated modern humans' major expansion from Africa. Converting a tiny part of that primal bounty into someone's experience of Shakespeare's writing seems like a fair trade, but it is also energy that might otherwise go on to do something else in the universe.

Energy is also flowing when any of those words are spoken to a rapt audience, or when tens of millions of dollars are spent to film their tales, or a TV is turned on to watch one of the plays performed, or we drive to a local Shakespeare festival. Or for that matter when we've impulsively bought a tacky bust of "the immortal bard" and hauled it onto a mantelpiece. Add in the energy expenditure of the actual manufacture of paper, books, tickets, tchotchkes and their transport, and the numbers only grow and grow.

Traditional paper remains an energy-hungry beast. In 2006, for example, it was estimated that US paper production alone gulped down about 2,400 trillion BTUs (about 2.5 million trillion joules) to churn out 99.5 million tons of pulp and paper products. That amounts to roughly 5 grams of high-quality coal being burned per page of paper, irrespective of whether it's a page describing a cure for cancer or a page of tawdry gossip in a magazine. This figure doesn't even account for the effort of forestry management of the trees that end up as paper, or the non-burdensome energy that it takes for a tree to grow.

The actual printed letters and words of a book typically use insoluble pigments rather than the dyes and organic solvents of pens or paints. Those other inks are made of delicious stuff like propylene glycol, propyl alcohol, toluene, or glyco-ethers, along with resins and wetting agents to give your ballpoint pen or felt marker the desired capabilities. Commercial printing pigments include compounds like carbon black made from the incomplete combustion of hydrocarbons, along with white titanium oxide for adjusting the hue—all of which have to be manufactured or refined. Globally, around 3.5 billion kilograms of printing ink are produced annually for a market that is today worth around US$40 billion.

Specific costs and energy use vary over time. But Shakespeare's plays alone, made into printed paper form over the centuries, may have quite easily slurped up the energy equivalent of combusting several million tons of high-quality coal. By comparison, a typical gasoline-run passenger car these days will end up emitting four to five tons of carbon dioxide during a single year. This means that you could also equate all of Shakespeare's printed plays with the planetary burden of using a couple million hydrocarbon-powered cars for a full year.

It may be impossible to ever fully and accurately gauge the total energetic demand that one William Shakespeare of Stratford-upon-Avon unwittingly dumped on the human species. But it is clearly substantial as individual contributions go. Of course, we can easily forgive him. He wrote some good stuff that has unarguably enriched the human psyche. But as we look at these numbers, and the myriad offshoots of objects and activities, I think it is impossible not to feel that Shakespeare's output has also taken on a burdensome life of its own, propagating itself into the future and compelling all of us to support it, just as is happening right now as I am composing—and you are reading—this sentence.

Shakespeare is, of course, just one particularly vivid example of how persistent and demanding data can be. He represents only a single drop in a vast ocean of seemingly ethereal human-made data that nonetheless has an extremely tangible effect upon us. That ocean, as we'll see, is both the glory and millstone of *Homo sapiens*.

It is also important to ask what would have happened if someone like Shakespeare had never existed. Would we have saved all that energy, or would we have expended even more energy doing other things? As my colleague, the astrophysicist Piet Hut of the Institute for Advanced Study in Princeton, has put it to me: instead of sitting at home quietly reading Will's plays and sipping cocoa, we might instead have gone to the local pub and started a fight. That would have surely expended far more biological energy and destroyed some barroom furniture as well. Extrapolating from this hypothetical example, one could argue that the net energy burden on humanity across the ages might actually be greater without Shakespeare than with him.

This clever observation raises the possibility that the energetic demands of having a dataome are misleading, since the alternatives could be even more demanding. The only way to sort this out is to evaluate whether or not those alternatives (including pub visits) might confer more benefits to our species than the dataome does. It is also possible that the phenomena of "us," of *Homo sapiens*, simply can't exist without its dataome. There might be no alternate reality in which a species like us persists without developing an external information system, along with all of its attendant burdens. Deciding exactly how these scales balance out is a substantial and challenging task that will take us on a deep dive through physics, biology, and the nature of evolution.

One thing that will become very clear, though, is that the overall growth in energy use by our species and the energy used in supporting the dataome track each other in time. That striking correlation suggests a strong, causative relationship between us and our information. A relationship that, as we'll eventually see, might literally take us to the ends of the universe.

## Entropy

Right now, by some accounts, our species generates about 2.5 quintillion bytes of electronic data a day. That's nearly a trillion times the data of all of Shakespeare's works being churned out every twenty-four hours. And the rate of output is still growing.

While lots of that data is interacted with fleetingly and is to some extent temporary—from Google searches and Snapchat postings to weather forecasts and air traffic control patterns—more and more of it ends up persisting in the dataome. That includes everything from movies and pet videos to amusing GIFs. Political diatribes and trolls' retorts, medical records, scientific results, business documents, industrial data, emails, tweets, and photo albums: all of that data winds up as semipermanent electrical blips in doped silicon chips or magnetic dots on hard drives scattered across the globe.

This data production and storage takes a lot of energy and attention to maintain and curate. Most of us don't often consider that energetic burden, but it's large. We tend to be shielded from this fact because by the time our modern computing and data storage devices and services get into our sticky little hands, their inner workings have been designed to be as invisible as possible. There is also a divide between the demand on resources to run our data hardware and the resources needed to build the hardware in the first place. The latter is perhaps the most obscure to most of us who simply plug things in and wait for miracles to happen.

A well-known study carried out back in 2002 by Eric Williams, Robert Ayres, and Miriam Heller discusses what they term the "1.7-kilogram microchip." This figure of mass comes from adding up all of the energy and chemicals employed during the production and eventual deployment of microelectronics. The authors estimated that a single, then-standard 32-megabyte DRAM memory chip represented an energy burden equivalent to fully combusting about 1.7 kilograms of refined fossil fuels and using about 72 grams of refined chemicals—all to build a fingernail-size device containing approximately 33 million microscopic transistors.

The researchers point out that this means a typical component of our data-rich world requires 630 times more mass in fossil fuels and raw materials than is in its final mass. That's a production mass ratio of 630 to 1. That can be compared with a production mass ratio of about 2 to 1 for the manufacture of the mechanical components and structural parts of an artifact such as a typical automobile. In other words, our machines for handling data are enormously more burdensome in a relative sense than other kinds of mechanical devices.

It's worth pausing to look a little harder at this. That's because the crazy-sounding difference between the demands of building a chip and a car can actually be explained by thinking about the fundamental physics of thermodynamics, and a universal property we call entropy.

Entropy will crop up again many times in this book. It is deeply, surprisingly, connected to the fundamental nature of information and to many cutting-edge insights on complexity and living systems. I'd argue that at its core, entropy is not mysterious at all, but, like many concepts in science, its original formulation was far from intuitive. That story begins in the physics developed in the mid-1800s during the culminating years of the Industrial Revolution. At that time entropy referred to the extent to which energy can actually make stuff happen in the world—can be *useful,* in the judgmental parlance of physics in the 1800s, a parlance that was heavily influenced by Calvinism.

In very general terms, the more entropy there is in a system (be it a heating system, a steam engine, or an entire planet), the less energy that system has available to do things like move a piston, or grow a flower, or print a book. And the really cruel thing that the nineteenth century physicists observed about the universe is that, on average, entropy is always increasing. This is one way to formulate the famous second law of thermodynamics, which has also been popularized as saying that not only can you never win, you can't even break even.

More specifically, anything in the world can be described as a system that is open, closed, or isolated in nature. An open system freely exchanges matter and energy with its surroundings, like a roaring bonfire. A closed system is akin to a heating system based on circulating hot water through pipes and radiators where energy, but not matter, can come and go between it and its surroundings. An isolated system cannot exchange anything with the rest of the world; no matter and no energy flows in or out, like the contents of a perfect thermos flask, or your metabolizing body packed into an entirely sealed space suit. The second law states, slightly more accurately, that the total entropy in an *isolated* system can never decrease. It might stay the same, or increase, but it can't be undone. This implies that there is an ever-worsening limit to the efficiency with which energy can be put to work in such a system; when energy is used (transformed), some of it will end up in unexploitable forms. The only way to beat this is when a system is *not* isolated, and can in effect trade entropy with something else. Although in the long term that's a bit of a cheat, because the net entropy of the system plus whatever environment it is in (like, you know, Earth or the entire cosmos) will still obey the second law.

In the case of machines and industry, entropy can be experimentally measured by examining properties like how vigorously a piston is driven by changes in the temperature and pressure of a gas inside it. But this can all seem murky, and luckily, by the late 1800s the concept of entropy had evolved into something more intriguing, and with far greater explanatory power. This was because a litany of famous scientists like the German physicist Rudolf Clausius and the developer of electromagnetic theory James Clerk Maxwell, as well as the Austrian scientist Ludwig Boltzmann, and the Yale physicist Josiah Gibbs, had developed the pieces of a field called statistical mechanics. It's an unfortunately dull name for a fantastically important set of ideas that use probability theory to relate the world's microscopic mechanics to its bulk properties, helping to explain what thermodynamics actually is.

Statistical mechanics fully embraces the true granular, atomic nature of matter. A gas, such as might be in an engine or piston, isn't just a gas anymore. Now it can be talked about as a *system* consisting of atoms and molecules whizzing around, and at any instant occupying discrete, specific things called *microstates* and *macrostates*.

Getting your head around what microstates and macrostates are takes a little sideways thinking: Imagine a giant mail room, full of sorting boxes or pigeonholes that can hold packages or letters. Now suppose that you arrange to store a bag full of letters in that mail room, which is overseen by a somewhat manic worker, whom we'll call the mail dervish. When your letters arrive at the mail room the mail dervish sorts them into some arrangement in the boxes. But, never satisfied with that arrangement, every moment the dervish re-sorts and rearranges the letters in a frenzied cloud of activity.

From your point of view outside the mail room, nothing has really changed. Your letters are all safely housed within the same-size mail room, with its pigeonholes or boxes. That external view is the *macrostate* of your letters. But at any given instant those letters can be in any one of an enormous number of possible configurations chosen by the frenetic mail dervish. Each of those individual configurations is a *microstate.* And each of those microstates—those arrangements of letters—can occur with differing probability, according to the whims of the mail dervish.

In the real world of gases and atoms and molecules, the letters are those atoms or molecules, and the pigeonholes or boxes represent all possible positions, velocities, and properties of the atoms or molecules. But when a cubic centimeter of air contains on the order of $10^{19}$ (ten billion billion) molecules, there are an awful lot of pigeonholes for a gas or any other substance. Statistical mechanics is the statistics of atoms or molecules in those pigeonholes, or their microstates. And statistical mechanics says that if the total energy of all the atoms or molecules is always the same, then the entropy is simply a measure of the total number of microstates.

That's extremely interesting because it means that entropy is also a measure of a certain kind of disorder. Suppose that the mail room actually only had a single box or pigeonhole. The mail dervish might be enormously frustrated by this, but it would have no choice but to place all the letters in a single configuration, a single microstate. This would be a mail room of low entropy, of order. By contrast, a mail room with thousands of boxes would offer a ridiculous number of options, and microstates, for the mail dervish to rearrange letters into at every passing instant.

That also means that arrangements of atoms or molecules (microstates) are far more likely to remain disordered than they are to spontaneously tidy themselves. That's our everyday experience of the world—how often has your dirty laundry picked itself up off of the floor or washed and folded itself? In principle, the quadrillions of atoms in your gym clothes and dirty socks could, by sheer chance, rearrange themselves into a clean and tidy configuration if you gave them a kick. But the odds of that single microstate are so infinitesimally small that you'd have to wait for what is effectively an infinite amount of time for this to happen. As Ludwig Boltzmann pointed out, there are vastly more future possibilities where disorder continues, where your socks stay dirty and scattered. Which is also precisely why more entropy results in less useful energy, because energy is soaked up in moving between disordered microstates that yield no work. All of this also implies that higher entropy increases our uncertainty about the specific microstate a system is in at any instant.

That's what helps make entropy so important: it's a gauge of both the state to which things tend to evolve (toward ever more dirty laundry) and of the probability that ordered, regular structure can even exist (how neatly folded you can ever expect your clothes to be). Restating this slightly differently: the inexorable growth of entropy has to do with the improbability that complex systems will ever retrace their steps to an earlier condition. As a consequence, to actually lower the entropy of any kind of object or system takes effort.

Which brings us all the way back to the observation that the mass ratio for raw ingredients and fuel needed to make a microchip versus a car are so discordant. Because of their small scale, computer chips and other high-tech artifacts are exceptionally organized forms of matter, with very low relative entropy. By comparison, a machine like a car has large components made of metal and plastics where the microscopic arrangement of atoms and molecules need not be particularly ordered at all.

Microprocessors and their ilk are also made out of what was initially much more disorganized, jumbled, and high-entropy material—stuff like craggy silicon grains and sloppy chemical mixtures. So, a microprocessor's exquisitely specific internal arrangements beat against the second law of thermodynamics and against the usual flow of entropy, and it takes an awful lot of work and energy to make them. To turn disorder into order in one little corner of the cosmos is a real headache.

This is a profound and important observation to make: Components of the human dataome face off nose-to-nose with fundamental physics. As we'll see later, both these components and life itself fall into a class of phenomena that temporarily violate the second law of thermodynamics. They are out-of-balance, out-of-equilibrium systems, sometimes far from the comforting expectations of what is most probable according to statistical mechanics. They are fluctuations, like a sudden blip or glitch in the world.

Another important example of how resources used by the modern technological world seem to push against the usual flow of things comes in the form of what are called rare earth elements. These might equally be called the "um" elements because their names are: scandium, yttrium, lanthanum, cerium, praseodymium, neodymium, promethium, samarium, europium, gadolinium, terbium, dysprosium, holmium, erbium, thulium, ytterbium and lutetium. If any group of elements in the periodic table can be called peculiar, it would be these.

The so-called lanthanide rare earths have atoms with an unusual electron arrangement that produce a host of special properties of light absorption and emission for these elements, as well as some impressive magnetic behaviors. Consequently, one major role for rare earths is in the magnets for highly efficient electric motors, such as those in computer hard drives. Other elements, like erbium, when incorporated into sections of optical fiber, can help with signal boosting. If you're reading this on a tablet or monitor, then yttrium may be in the white LEDs that are providing the main lighting source for your pixels. Lanthanum plays a role in high capacity batteries. Scandium aluminum nitride shows promise in semiconductor devices for handling high levels of electrical power.

The thing about the rare earths, though, is that while they are far from being underabundant in nature (for example, the element cerium is thought to be more abundant than copper is on Earth), their offbeat chemical behavior means that they're seldom, if ever, found in very high concentrations or seams. Instead, they're scattered in certain rock zones on Earth, and in certain kinds of hydrothermal systems—including some fragile and unique ecosystems at the bottom of the world's oceans. As a result, refining the rare earths is a costly headache. It's estimated that good, productive ores may contain only about 5 percent by mass of rare earth elements. Extracting these can involve lengthy processes that demand energy, cause environmental stress, and add yet more to the burden of the dataome's modern infrastructure. Importantly, were it not for the needs of human technology and data, it's not clear that these elements would ever get much use. No one would normally decide to make a saucepan or a sword from some terbium or erbium.

And rare earths are just some of the less-than-familiar elements that our world of information utilizes. Hafnium, a metal whose oxide has excellent electrical insulation properties and is important component of semiconductor devices, is often extracted as a by-product during the refinement of the element zirconium from natural sands—an extraction that involves a quite fearsome pyrometallurgical technique called the Kroll process, using molten magnesium or sodium. Another metal is tantalum, which is critical for making highly compact capacitors in our electronics. But one of the ores that tantalum is found in is considered a conflict resource in the Congo. It is not just costly to obtain, but costly for its role in war and human suffering.

## Electrons

Of course, the piece of the burden of the electronic dataome that we are most acutely aware of is the moment-by-moment use of electrical energy. It's a well-known developed-world headache to be constantly seeking the flow of current to power laptops and smartphones. That demand alone has led to a visible explosion of charger ports in public areas, from airports to coffee shops. But behind these seemingly trivial pieces of Earth's growing "technosphere," the challenges reach to an almost unimaginable scale.

Energy use in the United States has been well tracked, and estimated going back as far as the 1650s. By examining the production of wood, coal, oil, gas, hydroelectricity, and nuclear power, it's possible to reconstruct a reasonable approximation of the energy that one of the world's largest nations has been gobbling up across time.

The results are pretty amazing. Around the year 1650 the production rate of energy in the US already stood at about 100 million watts ($10^8$ watts). By 1800 it was topping about 10 billion watts, and by the year 2000 the US was producing some 3 trillion watts of power averaged out across a year. What that means is that since 1650 there has been a roughly 2.9 percent growth in energy production per year—without letup (intriguingly, it even increased a little faster during the Great Depression in the 1930s). This is exponential growth for most of the past 400 years.

Globally, a similar pattern emerges. In the early 1800s the planet-wide rate of energy production was on average around 100 billion watts. By the year 2000 it was in excess of a fearsome 17 trillion watts.

To understand the burdens and benefits of our informational world we need to know where the dataome fits into this global energy explosion. Recent estimates of the worldwide electricity use by data centers (those giant warehouses crammed with hard drives and servers supplying your pet videos and influencer episodes) suggest an average energy consumption rate of about 47 billion watts. There is evidence that this rate is still growing by about 40 percent a year. And this is solely the electrical energy, without accounting for manufacturing the storage hardware in the first place, or accounting for the personal devices that interface with it, from laptops to smartphones.

That means that our computational data centers alone are already using an amount of energy comparable to the entire planet's output in the early 1800s. If the present 40 percent annual growth were to continue, then in about eighteen years data centers would consume all the energy we currently produce as a species. As we'll see, this fearsome projection also tallies with the projections of computational efficiency improvements and the overall growth in our world of processors and computing.

Of course, we might argue that we're also getting better at producing electricity. Solar power is getting more efficient and vastly cheaper, and together with growth in grid storage (whether in batteries or stored hydroelectricity), it's feasible that electrical power will eventually become effectively free as we tap into more of the roughly 90,000 terawatts of solar power absorbed into Earth's surface. The issue is not that we necessarily run out of electricity, but that infrastructure for generating ever more power consumes ever more resources (in rare earths or lithium for batteries and their manufacture, even copper cabling, or sheer land area) and continues to feed technology that is itself utterly beholden to the second law of thermodynamics and the overall growth of entropy. Economically cheaper and more plentiful does not mean burden free.

A further illuminating comparison is with the typical metabolic power consumption of a resting human. That clocks in very approximately at around 65 watts. So supporting our large-scale computing resources is already equivalent (in a broad sense) to the biological energy needs of some 700 million humans.

If this seems inconceivable, one only has to look at the scale of modern data centers to feel a little uneasy. As I write these words, two of the very largest data centers in the world are located in Hohhot: the Inner Mongolia Information Park and the Hohhot Data Center. Hohhot is the capital of Inner Mongolia, part of the People's Republic of China, and its name means "Blue City."

Here at an altitude of over a thousand meters, with an annual mean temperature of about 43° Fahrenheit, air conditioning needs are—for now—far less than in many other places. Combined with readily available, and significant, geothermal and hydroelectric resources, this makes

Hohhot an obvious location to place extremely energy-hungry, temperature-sensitive, computational infrastructure.

The current champion in terms of size is China Telecom's Information Park, a complex of massive, modular buildings that clocks in at 10.7 million square feet of operating space, which is about 245 acres. Although this does include some living spaces and offices, the computer hardware capacity is astonishing. Right next to it is the current number-two size winner, the China Mobile Hohhot Data Center, covering 177 acres. Facilities like these can house tens of thousands of server racks—minimalized shelf-like electronic packages of processors and data storage, stack upon stack, row upon row. All gobbling power and reliant on a ceaseless flow of cold air to dissipate the heat of computation, to offload entropy.

Beyond the extraordinary thought that so much of the world's data is flowing through the former grasslands of Inner Mongolia, not so far from the birthplace of Genghis Khan, the size of these places is stunning. And they're not alone. Whether by building massive fields of solar panels in Nevada or slurping up hydroelectricity in Oregon, there are good reasons why a large company like Apple also maintains its own data server farms and looks for ways to optimize the power generation that these air-conditioned, electron-pushing factories demand. Although the power needs of microprocessors and electronic data storage have seen dramatic increases in efficiency, the simple truth is that we seem to be perpetually teetering on the brink of never quite doing well enough.

A study in 2016 by the Semiconductor Industry Association in the US produced a road map assessment of where things looked to be headed. The bleak prediction is the one I've already mentioned: by sometime around 2040 the world's computer chips will demand more electricity than is expected to be produced globally.

Behind this projection are issues that are genuinely tough to circumvent. The fundamental architecture of most current microchips involves vast numbers of interconnects—the wiring inside a chip, like the microscopic channels of conductive copper that hook together the billions of transistors in a processor. In a modern large-scale integrated circuit, the interconnects can total a length of thirty miles. All packed into a skinny little flake of a device.

But the smaller chips get, the more inefficient it becomes to push electrons around all of that increasingly narrow conductive material, and the more heat gets generated. More heat means less-functional chips, and a need for even more air conditioning or other forms of cooling. Combining those factors with the anticipated growth in the number of processors in use at any one time on the planet leads directly to the prediction that humanity's computing will suck up about a hundred quintillion ($10^{20}$) joules of energy per year by 2040. Compare that with the relatively paltry level of about a hundred trillion ($10^{14}$) joules per year today.

The implication is that if we want to avoid a crisis in the future we need to achieve many orders of magnitude of improvement in computational efficiency. Yet, other studies tell us that we're hitting a wall in devices' energy use per computation—the amount of energy needed to perform a simple procedure, like adding or multiplying two numbers.

Our shortfall in efficiency is brought into even starker relief if we also ask how far off we are from the computational efficiency of biological brains. A direct comparison is not that easy, but in 1998 researchers published a fascinating study on the metabolic cost of neural information, or what nerve cells have to do in terms of chemical energy production to launch their electrochemical signals. Their experimental subject was the lovely blowfly and its retina and associated neural cells.

What the researchers found was that a single bit of information—the on/off or 1 and 0 of binary logic—transmitted at a chemical synapse (the interface between neurons) involved about 10,000 of the common energy-carrying molecules called adenosine triphosphate, or ATP. For electrical spikes this rises to at least 1 million to 10 million ATPs. These levels seem to be significantly more "costly" than other cellular functions taking place, for example, in muscle tissue. Computation is not cheap even at a base biological level.

In pure energy terms, though, this translates to a single computation in a biological neuron costing around $10^{-15}$ to $10^{-14}$ joules of energy. That suggests the human brain may reach an efficiency as high as a billion billion computations per joule of expended energy. To put that in perspective: one joule of energy is about 0.00024 kilo-calories, or Calories, a unit we use to tell us the energy value of our food. In principle, just licking a bar of chocolate could provide enough energy to enable your brain to perform those billion billion basic computations (of course it's not quite that simple, but you get the idea). The bottom line is that biological brains are approximately a billion times *more efficient* than current standard microprocessor capabilities.

Possible ways to ratchet up the computational efficiency of our machines, to avoid going off a cliff of energy, include turning to radically different chip architectures. Some of these are directly inspired by the efficiency of biological neural systems, utilizing so-called neuromorphic chip design. But only time will tell if that kind of processor retooling will stave off the issue. And contrary to the popular notion that quantum computing is a panacea for our future computing needs, it is very far from clear that practical quantum computers will be better at everything than classical computers or more energy efficient. Indeed, the technical infrastructure needed to make quantum computation possible is truly formidable. While the computations themselves might use miniscule amounts of energy, creating the conditions for those computations and extracting their results is far from trivial. For now, when it comes to computation and data and energy use, we seem to be on a somewhat precarious and potentially unsustainable trajectory. We'll also discover, later in this book, that the universe itself sets a limit on the efficiency of any computation.

A final insult is the insidious, unavoidable, and fundamental aspect of physical reality that is our friend entropy. As we've seen, the way entropy evolves means that there is a limit to the efficiency with which energy can ever be put to work. Practically speaking, this intrinsic inefficiency is manifested as stuff like the waste heat of chip interconnects.

If waste heat is indistinguishable from the heat of the environment, it is useless for doing any more work. When we talk about energetic burdens, we're not just thinking about the supply of usable energy, we're thinking about what we'll never get back again: the loss due to entropy growth and our contribution to the increasing microscopic disorder of the cosmos.

You might still wonder if entropy growth is really what you experience in the world. After all, there are such things as rechargeable batteries that can be used again and again, and cardboard boxes that can most certainly be opened and shut again perfectly. The key is in the words I used when introducing entropy earlier: an isolated system. When something is not isolated you can decrease its entropy, but only by causing an inevitable increase somewhere else. Local violation of the law is possible, but it comes with some displaced cost. As we'll see, for us a part of the cost is that we are physically changing the planetary environment.

## Blockchain, Bitcoin, and Byzantines

Humans seem to be on a roll when it comes to finding new and entirely unexpected ways to burden our dataome and our computers. Ironically, we're often driven to this precisely because of the success of our data-oriented existence—more data in our lives increases the needs we impose on that data. One of the most dramatic recent examples comes from the esoteric and sometimes bewildering phenomena of cryptocurrencies and the scheme called blockchain. Digging into the mechanics of this is an eye-opening way to see the almost absurd demands that parts of the dataome, in turn, can make of us.

The basic idea of currency is as old as human trade. But unless you use truly rare, and therefore precious, materials like gold or silver to represent value, you run the risk of people counterfeiting worthless pieces of paper. To this day state treasuries go to great lengths to make their coins and notes extremely hard to replicate. The situation gets even more complicated online, where financial data needs to be uncopiable and encrypted, requiring special software "keys" to unlock it. Purely digital currencies, which can sidestep financial institutions altogether, take security even further, by decentralizing the underlying data across multitudes of separate computers and by making cryptography an integral part of the value of the currency itself. As a result, and as we'll see, these bizarre algorithmic constructions demand intense computational resources to validate any transactions.

Observed from a higher perch this is a fascinating development in the human dataome. Cryptography has been around a very long time, with devices like ciphers being used by many cultures thousands of years ago. But until the past century or so most of its computational load has been shouldered by human brains. Now it is encoded in our extended, external world. How that works is worth delving a little further into because it's an extreme example of the effort we're willing to expend to enforce the properties of information in our data, in this case the properties of rarity and security.

The most well-known of today's cryptocurrencies, Bitcoin, uses a record-keeping technique that is, in principle, inherently immune to retroactive alteration—the blockchain. The story of how the idea of a blockchain came to be, and the application to a currency like Bitcoin, is itself strangely hard to decrypt. For most of us in the world it began in October 2008 with the unexpected publication of a white paper—an unreviewed document laying out a technological schema.

That paper, "Bitcoin: A Peer-to-Peer Electronic Cash System," is a mere nine pages long and presents a secure model for electronic cash that, as the paper states, doesn't rely on trust. The proposed trick is to use a network of computers, a distributed system, and a scheme so that any malicious change to a public history of transactions (such as trying to steal or reuse currency) would require a completely impractical amount of computer time to cover its tracks. Just a few months later, in early 2009, the software to accomplish this was released, along with the first actual "Bitcoins."

None of this may sound particularly odd, except that the author of the white paper, one Satoshi Nakamoto, doesn't seem to actually exist. At very least this is a pseudonym, either of a single individual or of a group of people. Various claims have been made as to the true identity of Nakamoto, from known cryptography pioneers and computer scientists to economists and sociologists and even criminal masterminds. The whole tale is an entertaining mix of conspiracy, anti-establishment wishful thinking, and political dissent around the nature of international finance and economic governance. It's particularly fitting given the fact that a blockchain is a way to handle data in a world without implicit trust. If you feel like chasing the White Rabbit of *Alice in Wonderland* fame, then delving into this story is a great option.

But whether or not Nakamoto is real, the perceived value of a cryptocurrency like Bitcoin has, at points, been something to behold. As I write this a single Bitcoin is worth about US$10,000, and altogether cryptocurrencies have an on-paper value topping out at around US$300 billion. Little wonder that since 2009 so many people have ploughed resources into computer hardware that can be used to support the blockchain system and give them a piece of the action.

To understand what the blockchain really is, why it's a special change to the nature of the dataome, and why it is so energy hungry, it helps to look at one of the security issues that it solves. In a distributed computing or storage system, code or data is purposefully spread across many different physical locations, providing—in principle—much greater tolerance for failures or errors. It's like your backup has backups, which also have backups. But equally, to do anything useful with this distributed system, you have to be able to get at the correctly stored piece of information, meaning you have to be able to get a consensus result from those distributed machines that you can trust. If one of the pieces doesn't work properly, or if it's been tampered with, you need to be alerted to that fact, or better still, have the system itself deal with the problem. There's a name for this capability in computer science: it's called Byzantine fault tolerance.

Back in 1982 the computer scientists Leslie Lamport, Robert Shostak, and Marshall Pease published a research paper titled "The Byzantine Generals Problem." Previous debates on fault tolerance in computing systems had gone by labels like "The Chinese Generals Problem," or "The Albanian Generals Problem," but Lamport and his colleagues reasoned that "Byzantine" was less likely to be offensive to anyone living today. The story of the Byzantine Generals goes like this: There is a city under siege, surrounded by several Byzantine army camps, each commanded by a different general. The generals can only communicate with each other by sending out messengers, and they must coordinate their plans to attack the city. But some of the generals may be traitors, and the messengers might be captured and the messages tampered with en route. The question is how to ensure that the truly loyal generals commit to the same plan of action, and that a small number of traitors (if they exist) or interceptions don't cause the generals to adopt an ineffective, even disastrous, plan without realizing it.

Now, let's imagine that some of the loyal generals, after carefully studying the siege situation, decide that the coordinated attack should occur on Tuesday. We can also imagine that the generals are very savvy with cryptography. So, what they each do first is take the phrase "Attack on Tuesday" and add a random string of characters to it (in computer parlance, a hexadecimal value known as a "nonce"). The phrase might then become "Attack on Tuesday 5965732c205265616c6c79."

They then "hash" this phrase to generate a unique number. Hashing is a way to quickly read through and transform an arbitrary set of data, like a bunch of text, into a shorter string of digits. There are essentially an infinite number of choices for the rules that you deploy for doing this. A simple example would be that as you read through the text each letter causes a specific wheel in a rotating combination lock to turn by one step. So "A" always causes wheel #2 to rotate one step. "B" might always cause wheel #5 to rotate, and so on, according to the rules you decide on. The size of the hash result, perhaps a 10-digit number (like a ten-wheeled combination lock), is up to you to choose.

With a sizable enough hash result (those 10 digits or many more), any original message will generate a nearly unique code. Change or tamper with even a single letter in the original message and the hash result, the code, will be different.

That's a good start, and before laying siege to the city the generals all agreed on a hash function. But how do they validate the truth of any message? To do this the generals have also previously agreed that they will only send messages in which the hash result starts with (say) 6 zeroes. This is called a "hash target." To send a message they have to take their text, add a random string—the nonce—to it, and run the hashing process to see if the result starts with 6 zeroes. If it doesn't they repeat with a different added random string and so on until they get those starting zeroes.

If that sounds tedious, well, it is. And that's the most critical point. It takes time and (for our surprisingly technologically advanced Byzantines) a lot of computing power to assemble a message and its random string that produces the desired 6 zeroes at the start of the hash number. The effort required grows exponentially with the number of zeroes. This effort is called a "proof of work."

At this point the generals now have a way to generate a message that has its own check system. So, if the messenger hands over "Attack on Tuesday" together with its random addition or nonce, and any general who receives it runs this through the agreed-upon hash process and gets 6 zeroes at the start of the hash, they know the message was unaltered after sending.

If traitorous generals in the ranks, or city occupants, do intercept the messengers and alter the message, they will have to expend time on rehashing their altered message to get the magic 6 zeroes. So, you make their life hell by simply cranking up the computational requirements for generating a hashed message. As a result, they won't have time to mess with the communications.

This is the kind of solution put forward by Satoshi Nakamoto. The generals can hash together entire blocks of information, and demand 7,

8, 9, or more zeroes in the final hash target. Doing so means the computational demands of quickly changing—and therefore subverting—messages can be increased to the point of impossibility. In this case, even if a message is intercepted, it will take effectively forever to produce an undetectable alteration to it. And that's with just one messenger per general. Multiple messengers each with a piece of the final message would amplify the challenge even more. While even his process cannot ever be truly perfect, and indeed, if there are enough traitors around, or enough failures of process, it won't work, in most situations brute force is indeed what wins the siege.

That is, in essence, part of what happens in a blockchain. If you want to add data to the blockchain (or mint new Bitcoins, akin to unique messages), you have to behave like a loyal general, and put in the computational effort to solve a cryptographic puzzle like the hash process. This "mining" process is quickly verified by the network (the other generals who know what the hash target is). Illegally alter one piece of data and the hash won't match up across the network of computers. In essence, there's no way to tinker with the data without leaving your sticky fingerprints.

For Bitcoin and other cryptocurrencies, this means that the flow of currency can be governed by controlling the rate of introduction of new hash-like puzzles. But to actually create the currency, and to validate any given transaction, an immense amount of calculation has to be done to solve the distributed encryption. A lot of calculation means a lot of computers and a lot of specialized chips humming along. If money is involved (or at least the anticipation of future value in a currency), there is a huge incentive to buy up your own computer farms, devoted to mining the blockchain.

That's precisely what's happened thus far. Estimates place the annual power usage of Bitcoin at around 67 terawatt hours, or an average rate of energy consumption of around 7.7 gigawatts. That's about the same as the total output of a half-dozen commercial nuclear reactors. It gets even worse if you properly factor in the resources needed to make the necessary computer processors and support hardware. And that's just the most well-known cryptocurrency among many.

Currency, by its very nature, may have the most potential to drive the expanded use of blockchain technology and its energy-hungry implementations. But blockchains have many other uses too. In trade and commerce, the ability to securely track goods and transactions is vital. And for individuals, the use of personal blockchains offers an intriguing way for us to maintain control of our data, and our privacy. A company would have to request information about you from your blockchain—you would no longer exist as hundreds of data ghosts scattered across different databases. You would have control of your digital self and that digital self would be highly secure. Widespread identity theft would become an artifact of the early twenty-first century.

It seems likely that these kinds of computation-intensive technologies for ensuring truth in data, and personal control, will continue to grow as each of us places more and more of our existence into the dataome. In the larger perspective, against the backdrop of other life on Earth, this is pretty amazing. Our species is willing to put enormous effort into ensuring the integrity and utility of information, and utilizing the power of consensus. We're not just devoting physical computer time, we're devoting human thought. Whoever or whatever Satoshi Nakamoto is, a lot of biological neurons had to fire and grow in order to come up with the blockchain. Indeed, in an evolutionary sense it took centuries of human thought on the nature of calculation, automation, logic, and cryptography before blockchain had the opportunity to emerge in our dataome, and now it's swallowing up more and more resources.

## Machine Learning

Not only have a lot of human neurons spent time over the centuries thinking about new ways to create and engage with data, we've also looked for ways to offload those cognitive efforts. Today this is exemplified by the technology we call machine learning. Machine learning is a *subset* of artificial intelligence, or AI. It refers to any computational system that can perform tasks without explicit instructions, instead using patterns and statistical projections to make decisions, typically based on preexisting data or experience.

A broad subset of machine learning is deep learning, where the underlying mathematical algorithms are based around artificial neural networks with software architectures that, at least superficially, look like highly simplified brains. Virtual neurons are allowed to "talk" to other virtual neurons, from layer to layer of sheets of these virtual cells. Data moves through these networks, combining and correlating, feeding back on itself, and altering the properties of the virtual neurons. In effect the neural networks are never programmed; instead they actually learn how to program themselves. In that regard deep learning is amazingly powerful, and in many cases can produce computer systems that exceed human abilities at some cognitive tasks, from recognizing faces to acing video games. But deep learning systems are also complicated and messy. We don't fully understand why they work when they do.

Good machine learning for complex tasks takes a lot of processing power, because of the extremely repetitive, iterative methods needed to train a system like a deep learning one. By way of a simple example, something like the 2017 AlphaZero game-playing deep learning system and its descendants gets its smarts by playing millions of games with itself, or other versions of itself. AlphaZero's first outing also involved using over 5,000 specialized microprocessors. As a result, it became the most successful player of games like Go, chess, or shogi (Japanese chess) the world has ever seen.

A 2019 study by Emma Strubell, Ananya Ganesh, and Andrew McCallum at the University of Massachusetts, Amherst, examined the energy use and effective carbon footprint (the mass of carbon dioxide that ends up in the atmosphere) of training a number of large and sophisticated machine learning models. These models undertake demanding and intensive tasks like natural language processing, applied to translation and even to generating fake news. Thousands of physical microprocessors are typically deployed to run the codes. What the researchers revealed was that many of these machine training efforts produced enormous carbon footprints. The worst offender at the time, a Google-developed model called Transformer, often produced a footprint five times greater than the entire lifetime footprint of a gas-powered car (including its manufacture in the first place).

Yet these impressive (or depressing) costs are just a baseline. When researchers or companies develop new machine learning tools they end up performing multiple experiments, with many training iterations. Even when targeting more-run-of-the-mill tasks, this can add up to tens of thousands of pounds of carbon dioxide equivalent. Strubell and colleagues also found that some of the industry-standard tools for trying to incrementally improve machine training efficiency—by tuning or tweaking the virtual neural nets by trial and error—yielded relatively little benefit for a high cost. A few percent improvement might mean a lot in a competition against other commercial interests, but it's awfully expensive to attain.

Despite all of this it would be unfair to suggest that machine learning, and particularly deep learning systems, are an entirely hopeless situation in terms of energy efficiency. That's because the meteoric rise of machine learning has been driving new computer hardware development since the mid-2000s. Heavily used deep learning approaches, such as convolutional neural nets (multiple layers of virtual neurons, or perceptrons, that connect to previous layers and can, by brute force, learn the structure of complex data like imagery or video), carry out huge numbers of linear algebra operations on mathematical quantities called tensors. The mathematics used on these tensors involves a kind of giant matrix multiplication—capturing all the possible combinations of this multiplied by that, where the number of those combinations can reach astronomical sizes. Some of these computations are termed "multiply-adds" and completely dominate what a machine learning computer hardware system has to spend its time on.

Certain kinds of microprocessors are better disposed toward this kind of heavy-duty number crunching—particularly the graphics processing units (GPUs) that partner most personal computers and video game consoles. The manufacturers of these more specialized chips have created variants that include tensor-optimized elements, precisely because of the demand from deep learning. But there has also been a growth in fully specialized chips, such as the custom Tensor Processing Unit developed by Google. These advances all help, but even such efforts may not keep up with the future demands of algorithms for speed and energy efficiency.

The computer scientist Yann LeCun, one of the founders of the convolutional net approach, has suggested that not only will new chip architectures need to be better at basics like multiply-adds, they will also have to deal with much more complex and messy neural net arrangements. This is actually one area where quantum computing might help mitigate future energy costs. Work with enormous matrices is something that quantum computing should be able to handle superbly well, in principle.

But, as the famous physicist Niels Bohr once said "Prediction is very difficult, especially about the future." When I've talked to people who work in technology they will often argue that developments in computing and energy storage are quite likely to reduce many present inefficiencies and make what seems burdensome today a simple fact of life tomorrow. But while that has often happened in previous generations, the number of humans and the dataome seem to keep expanding to gobble up whatever gains we make. It's not clear that this can go on forever.

In 2019, the semiconductor manufacturer Intel launched a new kind of chip based on a neuromorphic design—more closely mimicking the interconnection of biological neurons. With some 8 million silicon "neurons," the chip was pronounced as being over 100 times more energy efficient for deep learning tasks. But, with some glee, Intel then explained that if it was scaled up by fifty times, as hoped, it would "only use thirty percent more power." The point being that higher efficiency merely enables greater computation, and continual growth in the actual net power consumption. With or without barriers of efficiency, the more we compute, and the larger the dataome grows, the more this burden will cause visible impacts.

## Noösphere Overgrowth

The idea that humans are severely altering the planetary environment out of which they evolved is not news to anyone today. But the notion that humans are a geological, planetary influence actually originated a long time ago. For example, in the 1920s an often-overlooked concept called the "Noösphere" appeared from a curious dialogue between a pair of French thinkers (a paleontologist and priest named Pierre Teilhard de Chardin, and a mathematician named Édouard Le Roy) and Vladimir Vernadsky, a Russian geologist. In its original form the term literally meant the "sphere of the mind." But Vernadsky, a serious and impressive scientist, was the most rigorous about formulating the idea, and in 1945 he wrote of the Noösphere as "a new geological phenomenon on our planet. In it for the first time, man becomes a *large-scale geological force.*" One particularly vivid example he uses is the fact that aluminum had never existed in its pure, metallic form on Earth until human minds figured out how to refine it.

This kind of re-sculpting, very much akin to what we do to support our dataome with paper, electronics, and machines, places demands on an environment and obeys certain rules. The result is unavoidable feedback between human energy use and the planetary environment. The fact that the burning of fossil fuels has cranked up atmospheric carbon dioxide to a level not seen in tens of millions of years is irrefutable (it's basic chemistry). That this has increased and will continue to increase globally averaged surface temperatures and alter the energetics and modes of climate and weather systems is also irrefutable (it's physics). That all of this will change human lives in a multitude of ways, many extremely negative, is also really without question.

In almost all instances and scenarios it will cost our species even more energy to cope with these changes, unless we transition to an entirely sustainable, zero-growth existence. Even making that transition would require, for a while, additional energy expenditures: manufacturing critical structures like solar panels and wind turbines, restructuring our distribution systems for food and fresh water, and building more energy-efficient housing. The cost of anything we do to the equilibrium of the planet tends to grow nonlinearly. Small costs at first can lead to much greater, and unexpected, costs. Even very modest global temperature changes result in statistically greater use of energy-hungry air-conditioning to keep humans and computers cool. For massive server farms that's a huge deal. Those same gains in global temperatures can shift climate zones, drying up the hydroelectric reservoirs that had seemed a sure thing for a sustainable future. There is an amplification of energy use in almost all scenarios.

If we'd paid attention to the physics of entropy and thermodynamics that we saw earlier, our species would understand all of this. If you recall, the original definition of entropy is as a measure of how much energy in a system is not capable of doing useful work. As entropy grows in a system there is less and less efficiency. Humans add entropy to the planet as they try to reduce the entropy of their lives and their low-entropy technology. And there's a limit to how much Earth can offload that entropy gain to the rest of the cosmos over a short period.

So just how much of total human energy use (and entropy generation) is directly attributable to the dataome? Even though I've sketched out some of the dataome's burden, its relative contribution is extremely hard to pin down with perfect confidence. To do that, we'd need to distinguish energy use solely in service of the dataome from other uses. Arguably, though, there is no real separation. The energy required to build a house is certainly in service of a basic human need for shelter. But houses often have architectural plans, measurements, and functionality in terms of protecting not just fleshy humans but also their documents, books, TVs, and computers. The machines we use to forge and assemble the materials are also only possible with a dataome to store their complex blueprints and manufacturing details. The dataome has insinuated itself into our most basic survival needs. As a result, the exponential growth in civilization's energy use and in the use of energy for data—from books to machine learning—may well be one and the same thing.

The interdisciplinary scientist and writer Vaclav Smil has (most recently in 2019) scoured data on human growth and energy use throughout history to try to envisage what the future holds. It's not a pretty picture. While the raw economic cost of computing has fallen by about 100 billion times since the days of vacuum tubes in the last century, it's hard to argue that's resulted in less burden on the world, a fact we see in the figures on energy use I've presented. Even agriculture, although capable of extracting 10 times more food energy per unit area of land than a century ago, accomplishes this because of a factor-of-90 increase in total energy input. That energy is transferred from hydrocarbon combusting machines, electricity, and technology, and is ultimately routed through the dataome on its way to fields of soybeans or rice.

From books to computers to Bitcoins and machine learning, the modern dataome represents an eye-popping burden. But the burden described solely by the direct use of energy is too simple a story. There are also complex burdens tied to the way we live, work, and behave.

## Obey the Data

In 1889, a mechanical engineer named Herman Hollerith persuaded the United States government to use machine-readable paper punch cards and electric counting machines to conduct the national census, saving the US Treasury about $5 million in the costs of handling data for 62 million people. By 1891 Hollerith was designing machines to be used in censuses in Canada, Norway, and Austria, as well as by railroads to tabulate fare information. As a result, Hollerith's fledgling company quickly became part of the growing data landscape, and eventually, in 1924, it became the International Business Machines Corporation, or IBM.

The core technological idea that Hollerith exploited had originated in the early 1700s with punched holes in paper tape that could govern the movement of thread in automated weaving looms. This mechanical hack meant that complex patterns could be reproduced quickly and efficiently in all kinds of textiles. And during the 1800s punched paper grew beyond looms into a more general technique for information storage and manipulation.

By 1937, IBM was churning out an astonishing 10 million blank punch cards per day. In the first half of the twentieth century, these simple paper cards, designed to take a precise grid of punches (literally rectangular holes) of 80 columns by 12 rows, were used in an enormous variety of business and governmental applications, from accounting records to personnel files. Their tiny papery voids were data manifested in the physical world.

The twentieth-century dataome was, for a while, drowning in punch cards. In World War II the cryptographers working at Bletchley Park in England would routinely use about 2 million punch cards a week to store their decryptions of Nazi communications. But that wasn't the peak: that came in 1967, at which point the US alone consumed approximately 200 *billion* cards per year. By then punch cards were also the common medium for programming and interrogating fully digital computers.

Extremely critical tasks like the core programming of NASA's Apollo project's Guidance Computer, whose design was overseen by the computer scientist Margaret Hamilton as director of software engineering at MIT's Instrumentation Laboratory, were transposed onto IBM paper punch cards. This process was entirely standard routine for software development. A programmer would write their code by hand on special coding sheets—preformatted forms. Those coding sheets would then be converted into physical punch cards using a keypunch machine, essentially a typewriter attached to a powered machine equipped to punch the holes in the cards. Often this procedure would take place twice, where a second human operator would punch the cards again in a verifier machine that compared this new set with the first to catch any errors. Finally, the cards would be fed to a card-reading machine that would convert the holes to electrical signals, physically compiling the data into the electronic computer.

All of this meant that using a computer was a physical chore—from making the cards to carrying a stack of them to the card reader and waiting your turn to program the computer. In software parlance this was a batch-mode process with you as a part of the batch. And as with all of our books and printed materials, punch cards also represented a significant energy burden. Factoring in the paper production, the transport, the machines, the human effort, it's likely that in the late 1960s an equivalent of somewhere between 1 percent and 10 percent of the national coal-burning energy budget in the US was going into the production and use of punch cards.

The fascinating thing is that punch cards, perhaps even more than ordinary printed matter, demanded very specific human behaviors. The whole thought process of writing, testing, and using computer software was, for a while, beholden to the quirks of a format that had its origins in how we mass-produced pretty textiles. There was also a direct physical relationship to the data as stacks of cards were carried and deposited, organized and discarded. For a while in the 1960s it must have seemed like our future was truly that of a cog in a giant electromechanical organism.

But by the mid-1980s, punch cards went pretty much extinct. Even IBM wasn't making them anymore. Instead, the cards and the machines that used them became relegated to specific legacy needs. Magnetic tapes and disks and solid-state computer memory became the norm. These new media were vastly more compact, faster, and more reliable. Critically, these storage techniques also lent themselves to a much more natural approach to data. Imagine trying to record and digitize an orchestral performance with punch cards. Not only would this be spectacularly slow, the machinery would need to be highly customized. By contrast, that music can be almost seamlessly and finely digitized and recorded onto magnetic tape or a silicon memory—with an enormously more accommodating set of limitations on speed and capacity. The barrier between the dataome and the physical world, and us, was magically lowered. Punch cards enabled the rise of the digital world, but they were also outdone by it.

There is a striking similarity between the history of punch cards and the kinds of patterns we find in evolutionary biology and the ebb and flow of species. That is not necessarily a profound thing. A correlation and similarity do not imply an underlying fundamental linkage. But seen through this lens it's pretty undeniable that these simple pieces of paper take on an organism-like character, in an essential symbiosis with humans.

Punch cards were initially an emergent innovation in a niche environment (of mechanical looms and tabulating machines). But as their utility was exploited for a broader range of purposes, they in turn influenced the environment. In biological terms, which we've already encountered, their phenotype extended.

They placed demands on energy and raw materials and sculpted their surrounding infrastructure. They also reached back to directly govern human behavior—an interpretation that I will argue makes sense to take quite literally. The punch card component of an evolving dataome resulted in humans training themselves to punch the cards, to carry them to the machines, to painstakingly disentangle things when stuff caught and jammed. To live, think, and dream of punch cards.

Special printing presses needed to be invented to produce these cards, and engineers worried over excruciating minutiae like the shape of the punches—circular versus the more efficiently packed rectangles. The specific data passing through the cards at any given moment, whether the US census or an Apollo lunar trajectory, was almost irrelevant. In the dataome, the punch card was an evolutionary experiment, a new kind of biomechanical structure if you will. In the process, it changed the ways we think about information and the things we can do with it. Punch cards helped drive human society out of the industrial age and into the data age like an unruly herd.

Eventually they were selected out of existence because they were too energy hungry and too cumbersome. Arguably, though, some of their "genes" survived through this extinction. The way that these cards structured data included designating years with two digits—to save space. This feature created the threat of a so-called "Y2K" disaster as the human calendar passed the year 2000. The nomenclature used by some computer languages for formatting data also has its roots in the need for standardizing punches in their 80 columns. These traits have indeed outlived their material hosts, and carried a little bit of the burden with them.

It's all too easy to assume that today we're done with such temporary and demanding phenomena as punch cards. But of course that isn't necessarily the case. If I was writing this from fifty years in the future I might be rolling my eyes at the craziness of cryptocurrencies of the early twenty-first century. Or the lunatic naïvety of smartphones and social media and how they made us behave. There is no reason to suppose that other novel and differently burdensome experiments won't keep emerging from the dataome and supplanting their predecessors.

## Burden to Benefit

Why are we really doing any of this? Why are we expending ever-increasing amounts of effort to maintain and use the data we, and our machines, generate? This behavior extends far beyond the oral traditions of ancient cultures or the cuneiform imprints of a few thousand years ago. I think that the reasons run far deeper than our appetite for literature or binge-watching. Indeed, they may explain why such things appeal to us in the first place.

The easy, almost lazy hypothesis is that our capacity to carry so much data with us through time is a critical part of our success at spreading across the planet. The data we maintain outside of our biological selves seems to provide us with a massive evolutionary advantage. We can continually develop our knowledge and experience in a way that no other species apparently does. The dataome may be, in very real terms, the grander scaffolding for complex, technological life.

That might sound self-evident. We like to consider ourselves "smarter" than other species, and those smarts let us manipulate our world, bend it to our bidding. Our informational baggage is just an inevitable part of that. But evolution is a tricky business. Its rules, as we'll see, are not careful or considerate, and our pact with the dataome is Faustian in nature. Every idea is a burden in some way or other, whether in biological metabolism or technological energy demands, but not every idea is equal in its value or utility: Is that tune that you can't stop humming since you heard it in a TV commercial really equivalent to the idea you just had for reversing climate change?

To understand what the real benefits are from a dataome, we have to examine what's in ours and consider whether all data are born equal or not. We have to travel a lot further into the darkening thickets of data, information, and machines.

# 3

# IN SICKNESS AND IN HEALTH

*The greatest value of a picture is when it forces us to notice what we never expected to see.*

—John W. Tukey, 1977, *Exploratory Data Analysis*

Near the town of Maros in the hilly, tropical forest landscape on the southern end of the island of Sulawesi in Indonesia are hundreds of stalactite- and stalagmite-filled caves eaten out of the soluble limestone rock by eons of watery erosion. In at least ninety of these subterranean cavities are the remains of a distant era of humans. There are prehistoric stone tools, the remnants of harvested shellfish. And there are paintings.

Stylized, even abstract versions of creatures like pigs and dwarf buffalo adorn the walls in various states of preservation. They are eerily beautiful representations of a world that existed as far back as 40,000 years before today. But the most dreamlike and moving works of art of all are the hand stencils.

Tens of thousands of years ago, and at many different times since, humans held their forearms up against the rock and sprayed—likely using their mouths—a watery mix of the pigment ocher across hand and wall, making shadowed outlines of their five fingers. Sometimes these stencils are an isolated statement, but more often than not they come as clusters; crowds of ghostly hands splattered across the rock and overlaid on each other.

In other places they are embedded within the depictions of animals, an expression of a relationship between human and beast that is simultaneously impossible to divine yet immediately familiar. Our gut reaction is that of course someone would do that; it feels magical, powerful, and aesthetically satisfying even if we can't say exactly why. The hand stencils are extremely personal but very alien remains, intact in size, shape, and pose. You can place your own fingers in precisely the same place as another human being from so far in the past they might as well come from another world.

These are not the only locations where hand stencils have been discovered. Good examples appear on cave and outcrop walls in Spain, France, Australia, Papua New Guinea, Argentina, Turkey, and South Africa, spanning times from at least 40,000 years ago to a few thousand years ago. Intriguingly, there is evidence that some stencils in Spanish caves even predate the European arrival of *Homo sapiens*. Instead, over 65,000 years ago *Homo neanderthalensis* were flexing their own cognitive and artistic skills and leaving their marks. We'll come back to that startling proposition a little later.

Regardless of the minds behind them, in most of these cases the hand has been used as a repeatable pattern that is accurate and efficient. It's instantly recognizable, and if humans in the twenty-first century feel a connection and sense of familiarity, the chances are so did our ancestors across history. The information conveyed may be a little ambiguous or variable depending on the artist, but it has immense *meaning*. At the same time, the storage system is nigh on perfect in terms of reproducibility and fidelity. If you want an example of low-burden, high-information content data, these ancient hand stencils are pretty healthy-looking. They place little to no energetic demand on us, require little immediate decoding, yet contain a wealth of potential insights into human history and intellectual development.

But if I can casually use the adjective "healthy" to describe information, can I also use the word "sick"? And what do I mean by "meaning"? In the English language sickness is defined as ill health or a disordered, weakened, unsound condition. Sometimes it refers to the consequences of a particular disease or malady that negatively affects the function or structure of an organism or any of its parts. In the natural world sickness tends to be easy to identify. A tree might look rotten and drooping, stunted and yellow. Animals are sluggish, poorly, or dead. But what does sick data look like? The most obvious example is when data is stored but corrupted. Most of us have had that sinking feeling when our computer screens blink with the news that a file is unopenable or damaged. Yet there it sits, still taking up in-silico space. Or maybe we've dealt with water-damaged books, their pages of cherished words blurred by diluted ink or turned to mush as the paper unbinds.

Like those sodden books, truly corrupted data tends to be removed forever from the dataome. It is winnowed out by a kind of fitness selection—if it can't propagate or fulfill some informational purpose, it dies. A much trickier proposition is data containing information that seeks to deceive, or that has no correspondence to real phenomena. Or data that has low utility, or high levels of duplication, or is simply full of random inaccuracies. These are all truly murky areas, because the utility of data and the information it contains may not always be immediately apparent. I doubt anyone who created the images of their stenciled hands on cave walls could have possibly guessed their scientific worth centuries later.

There are striking examples in nature of what happens when duplicated, biased, or deceptive data is unleashed. Army ants have minimal visual sensitivity but exquisite olfactory and touch senses. Like many other species of ants, they secrete their own chemical navigational and communication aids in the form of pheromone trails—a rudimentary, albeit temporary, externalization of information. If an ant finds a trail it tends to follow it to food or other members of its colony, or army. As it does so, it will add more pheromones to that trail, strengthening it. In most circumstances this is a brilliant algorithmic solution to complex problems. If a trail leads to food, no matter how convoluted it is, the back-and-forth of ants to the supply will automatically strengthen that path as being successful. But if an ant, in exploring, circles back onto its own trail, it can be deceived and start following itself in an endless cycle that just further amplifies the mistake.

For army ants swarming across the landscape, this can result in a spectacular phenomenon called a spiral of death, or in more muted terms a "mill": hundreds, if not thousands of ants swirling around and around, following their ever more potent chemical trail. The largest reported instances of army ant mills have been over a thousand feet across, with the outermost ants taking upwards of two hours to complete their pointless circuit. Unless something breaks the pattern, such as a falling branch or a new nearby food source, the ants can circle until they die. That is, by any rational standard, a very dire consequence of unhealthy data and misleading information.

## Big Data

Today we can look at the torrent of data from the daily online behavior of the three and a half billion people with broadband internet access, and wonder what use most of it is and what its real meaning is. Cynically, we might also wonder if, like the unfortunate army ants, sometimes we're simply following our own data trails around in circles. There are the websites you repeatedly skim across in case something new has appeared in the last two minutes, the junk email you forget to delete and tag as spam, the emoji-infested text messages you feel obliged to respond in kind to (happy face, thumbs up, heart). The advertisements for human-size hamster wheels and homeopathic suppositories that appear on your screen because you already foolishly looked at similar advertisements.

There is also an ocean of data you may be entirely unaware you're sloshing around in. When you use public transport with an electronic payment, a few bytes are being added to the dataome. Ordering that complicated coffee or a fancy sandwich will often involve an electronic till. Even if you pay with cash, a database somewhere is being updated. Buying groceries can involve scanning bar codes and triggering a chain of data products from sales records to inventory and automated warehouse restocking requests. Simply walking down the street can involve your capture onto security camera hard drives and cloud computing servers on another part of the planet.

Entire swaths of modern industry are built around the idea that there are uses for all of this data and more information to be squeezed from a seemingly mundane sponge. The specific label is "Big Data," a name that has been around since the 1990s (and ironically that label has origins that are hard to pin down; the data is imperfect). A central proposition is that there will be patterns and structures in very large datasets that can be useful for understanding, for example, why people behave the way they behave or what they like and don't like, even if they don't know it. There might also be hidden phenomena: trends and patterns that can help predict the future, whether it's for the popularity of lattes or the price of airline tickets to Lapland.

There are also things to be learned from big data that could genuinely help our species—arguably providing meaning because they aid our continued existence. Centuries of medical records have the potential to reveal new correlations between behavior, heritage, environment, and health. Country-by-country and global data on people and economies can provide critical help with a nation's development by supporting decision making on governance, finance, and more. Real-time data on everything from traffic patterns to internet searches can reveal the beginnings of epidemics or the migration of vital insect species and the human-created dangers they face.

But there is no guarantee of hidden gems in big data. There is also no guarantee that the knowledge gained by sifting through enormous amounts of information will compensate for the energy expended to do so. And there is a fundamental problem today with big data about human lives. That is due to data inequality—the "digital divide." According to figures from the World Bank, more than 3.8 billion people do not have easy access to the internet. That's half the current human population and half of the entirety of human experience that doesn't get to interact with the dataome as the other half does. Worse, 20 percent of adults are still functionally illiterate, drastically limiting their connection to the dataome as well. There are two things implied by all of this: First, any big data that is meant to reflect the properties of human populations is going to be intrinsically skewed. Second, not all humans are exposed to the potential benefits of the dataome, yet they are going to experience the effects that it has on other people and on the planet. As a result, grading the genuine utility of our data—big and small—is not at all straightforward.

Then there is data that we generate almost precisely because it doesn't seem to have any practical utility. Social media is crammed with this stuff. Amusing pet videos or snippets of accidental mishaps overflow, as do animated GIFs and selfie after selfie, or images of your most recent meal. It may be that in the future all of this will be seen as a vital resource, but that future may be a long way off.

It doesn't seem like too much of a stretch to claim that a lot of our data, especially in electronic media, is of intrinsically low utility to our species as a global whole. In the worst cases it causes us to spin in pointless circles. That doesn't just place an unnecessary burden on our energy infrastructure; it actually creates a hurdle for getting at data that has genuine utility.

It would be easy to characterize this as sickness in the human dataome. The catch is that we don't know whether this is something out of control or in dynamic balance with all the benefits the dataome confers on us (which we'll explore shortly). Asking whether all of these dataome properties and pitfalls are a predictable outcome, determined by the fundamental nature of information and the organisms that generate and curate it, will take us into some very strange but important territory. We'll become witnesses to a slow-burning scientific revolution that changes what we think data, information, and meaning really are.

## Infor . . . ion

Intuitively we might guess that there has to be a rigorous way to evaluate data's utility and gauge its core value. It must be possible to figure out how much information there really is in something. After all, I can immediately judge that

> "ddddddddddddddddddddddddddddddddddd"

has less to tell the world than

> "Now is the winter of our discontent"

—even though both take up an equal number of character spaces on the page.

It turns out there is a beautiful way to mathematically evaluate this difference, at least in a very austere, literal fashion. That mathematical evaluation is the first part of an ongoing and fascinating scientific story, one in which information carries meaning, and meaning depends on layers of context, knowledge, and *how* information leads to actions and decisions. As we'll see, information turns out to be an epistemic property of data, as a measure of knowledge and knowability.

The surprising genesis of the mathematics of information resulted from the ever-increasing demands being placed on electronic communications around the world in the early 1900s, from the basic telegraph and its thousands of miles of often rickety wires to radio and its fluctuating but potentially limitless reach. Two important and linked challenges emerged from these demands, challenges that still drive research today. The first is how to make data transmission as reliable as possible. The second is how to squeeze a set of data down to its minimum necessary description: the most compact and efficient state that preserves all, or most, of the information it contains. Neither of these tasks has an obvious, intuitive answer.

Let's start with that squeezing of data. A very fundamental measure of information content comes from asking how compressible a set of data is. Or, more specifically, how compressible data is without losing any information. That may seem confusing, so let me elaborate. For most of us data compression is about making a file take up less space on our computers, smartphones, or memory drives. Your favorite online movies or shows are likely to already be in a compressed video format so that they download or stream faster. Filenames ending in codes like "jpeg" or "gif" also correspond to images that have been shrunk by some amount in storage size from their original raw form straight from a camera's sensor. Compression is about eliminating extraneous information but preserving, to a variable degree, the important parts of data. The trick is to know what's important, or meaningful.

A simple thought experiment helps to frame this. Imagine a checkerboard of black and white squares. What's the greatest compression I can apply to this mental picture without losing anything? The answer is in the name. If I tell you "it's a checkerboard" the chances are good that you'd be able to reconstruct this object accurately. In this case your prior knowledge allows an extraordinary degree of data compression, down to the single word "checkerboard." About the only extra things you might need to have specified are what the corner color is and how many squares there are. This is an example, of a sort, of "lossless" compression where nothing important is left out. But part of why this works is that the checkered pattern matters to *us*: we utilize that pattern to play games. It has a knowledge-based meaning and already has an informational presence in our mental world.

Now compare the checkerboard description with "It's the *Mona Lisa*." You'll know what that is, you may even have a good idea of the portrait's form and some of the details, but unless you are a reincarnation of Leonardo da Vinci, you won't be able to reproduce it very well. You and others will also differ on what is most important in this painting. This is an example of very "lossy" compression.

In principle it is possible to compress the information contained in the *Mona Lisa* to some degree without losing anything that matters to anyone (within reason). For example, in the painting there are areas of nearly uniform coloration and intensity: parts of the subject's skin, areas of her clothing, and sections of the background landscape and sky. Those regions really do contain less information per unit area and could be described as outlined shapes using much less data. Even the painting's overlaying patina of fine cracks could be described without reference to every microscopic speck of pigment. And to the human eye and brain the loss of a few percent of the fine details of the Mona Lisa may be imperceptible. The utility of the painting is not hugely compromised by a small amount of loss in compression.

Nonetheless, the absolute compressibility of the checkerboard versus the *Mona Lisa*—whether lossless or a little lossy—is very different. We can intuitively see that the checkerboard contains far less information than Leonardo's masterpiece. In other words, compressibility is one way to evaluate how much information is really contained in a dataset.

But what tells us, mathematically, the true level of lossless compression for some data? It turns out there is an even more general way to look at all of this, involving communications, our old friend entropy, and the field of information theory.

## Telegraph

At its roots, information theory is a fancy way of talking about data transmission, and how to accomplish that with efficiency and accuracy in a noisy world. Here "noise" has a very specific meaning: the unintended and unpredictable fluctuations that can distort or complicate a signal, whether in a wire, a flashing light, or a radio wave. The very first transatlantic telegraph cable in 1858 had such bad, noisy reception that it took two minutes of repeated Morse code to send a single recognizable character, and consequently an astonishing 17 hours and 40 minutes to send the first official message across an ocean. That message consisted of ninety-nine diplomatically choice words from Queen Victoria in London to President Buchanan in Washington, DC.

We've all dealt with versions of this problem at some point or other in our lives. Say you see your friend across a busy street and you want to grab the chance to catch up. What you'd like to do is carry out a conversation like: "Hello, Alice, nice to see you. It's been a while. How is everything going at the new job?" But, aside from the social faux pas of yelling that out across the thoroughfare, if you did, chances are that Alice would only be able to understand fragments because of all the background noise added to your voice.

Your instinct might be to simply raise the volume of your voice. But as we all know, there are limits to that, and it becomes increasingly challenging to summon enough lung power if the street gets noisier. For other means of communication, like making a transatlantic phone call in the 1920s, simply boosting the power quickly became impractical. Indeed, that first transatlantic cable back in 1858 ultimately failed in part because an equipment-damaging 2,000 volts were applied in a bid to improve things. There had to be other ways to overcome these limits.

One place where scientists first tackled problems of this ilk was at Bell Laboratories in the United States. The Labs, part of a communications empire that extended back to Alexander Graham Bell, were a hotbed of extraordinary engineering and scientific prowess. They were crammed with innovative thinkers encouraged to spend their time thinking (sadly, mostly white men at that time, something that only changed—dramatically—in the 1970s). It was here where these seemingly mundane questions of communication transformed into fundamental puzzles about the very nature of information.

An important moment came in 1924, when the electronic engineer Harry Nyquist developed a way to calculate the maximum number of signal pulses (informational blips, beeps, squeaks, whatever you want to call them) that could be theoretically sent down a telegraph wire in a given time. That rate of signaling turned out to be exactly twice the system's bandwidth (the range of frequencies it could handle). There was simply no point in trying to send information faster than this.

This doesn't sound very exciting, but it represented a profound shift in thinking. It meant that data and its information was, in some yet-to-be-figured-out way, dependent on the flexibility of choice you had in your symbolic signals. The actual contents of a communication were therefore inextricably linked to a system's measured limitations (the bandwidth of a telegraph wire, for instance). Researchers were starting to see information as a quantifiable substance.

Then in 1928 Nyquist's Bell Labs colleague Ralph Hartley produced a paper called simply, "Transmission of Information." In his own words, the paper presented "A quantitative measure of 'information' . . . which is based on physical as contrasted with psychological considerations." This too represented a pretty radical change in how to approach the whole notion of information. It ditched the human interpretation, the psychology embedded in what is or isn't recognizable as information, and instead thought of information as something more absolute, with an independent existence.

A key piece of Hartley's work was the idea that constructing any message, in any language or abstract representation, involves selecting specific symbols out of a larger number of possible symbols. But as you select these symbols you effectively eliminate others from being used. Hartley himself used the example of the sentence "Apples are red." The first word effectively eliminates any other kind of fruit, as well as any other objects. The second word homes in on a property or condition, and the third excludes all other possible colors.

The upshot is that the value of information (how informative the information is) seems to be related to how much each piece of it eliminates other possibilities—how many alternatives are now excluded. This is very much like the ordinary game of "twenty questions," where players attempt to guess an object by asking questions that systematically eliminate ranges of possibilities. The best, most informative questions are the ones that get rid of the most options. Critically, this informational value also relates to the number of possible symbols that could have been chosen. If you pick a symbol out of a large vocabulary you might eliminate far more options than if you had a small vocabulary. That means, in these newly invented terms, that your chosen symbol is more informative.

All of these studies at the Bell Labs laid the groundwork for someone to come along and pull the pieces together, and so someone did. In the history of information theory, and science in general, one of the most influential research papers of the twentieth century is Claude Shannon's "A Mathematical Theory of Communication," published in 1948.

In forty-five pages of not-for-the-faint-of-heart calculations and demonstrations, Shannon laid out the basis for determining all of the fundamental limits of signal processing, communication, and information. Or, in plainer words, how efficiently and accurately you can ever hope to talk to your friend Alice across a noisy street, and clues for how to do that.

Shannon was a special kind of character; brilliant, colorful, and a little strange. Born in 1916 in Michigan, he graduated university at age twenty with two degrees, one in electrical engineering and one in mathematics. During graduate school at MIT he produced foundational work on how to use switching circuits to perform what we call Boolean algebra—the algebra of true and false, 1 or 0, and logical operations: the seeds of what would become the logic basis of modern computing.

By the early 1940s he too was employed at the famous Bell Labs. During this time Shannon worked on many things, including cryptography. He also got to interact, albeit briefly, with the mathematician Alan Turing, who had been posted to Washington, DC, for a while in 1943. Turing, whom we'll encounter again, is often considered the father of modern computer science. Shannon also became known, or perhaps notorious, for his proclivity for juggling and riding a unicycle (he reportedly had upwards of thirty unicycles in his garage later in life, including one with a square wheel), and later inventing and building a flame-throwing trumpet, the purpose of which remains obscure.

But when it came to examining the nature of information, Shannon's motivation was, at least initially, fairly mundane. Like Nyquist and Hartley in the 1920s, he wanted to better understand the pros and cons of specific engineering solutions to signal communication, and how to best encode messages to maximize the chances of their accurate reception. But this work led Shannon to a vastly more general theory, one that finally made information or data into a tangible "thing" in ways that it really hadn't been before—somewhat to his chagrin.

What Shannon added to earlier insights was that it isn't just the size of the symbol vocabulary that matters, it's the *probability* that a particular symbol will be chosen. Information is inherently probabilistic. It is, to paraphrase Shannon, a measure of the reduction in uncertainty we achieve, or our *surprise.*

Imagine that you toss a coin and 50 percent of the time it comes up heads, and 50 percent of the time tails. That means that each time you toss the coin there's maximum surprise—you simply cannot predict the outcome. In that sense you *learn* the most you can with each toss; the coin provides maximum information. By contrast, a heavily biased coin may come up heads 100 percent of the time. When you toss that coin there is absolutely no surprise, and you also learn absolutely nothing new each time you toss it. The biased coin's information content is zero.

The coin is also an example of a binary system. One side can represent 1 and the other 0. An unbiased coin stores one "binary information digit." The term comes from one of Shannon's colleagues, the mathematician John Tukey, but Shannon, in his 1948 paper, shortened it to the now ubiquitous "bit." By contrast, a biased coin effectively stores *less* than one bit.

So far so good. But humans use languages, and languages have complex and not always logical rules that dictate how we use symbols and collections of symbols. Shannon took this thorny problem on in order to illustrate his theory. He did this with a lengthy but magical example that I reproduce here because, well, because it's cool.

His first step was to take a huge printed table of random numbers (this being the 1940s) and plunk his finger down at random places. Where his finger pointed he'd look for a number between 1 and 27—corresponding to the 26 letters of the English alphabet plus a space. Then he'd do that all over again. In effect he rolled and re-rolled a 27-sided die to create a truly random string of letter "noise." Here is his actual first attempt:

> XFOML RXKHRJFFJUJ ZLPWCFWKCYJ FFJEYVKCQS-
> GHYD QPAAMKBZAACIBZLHJQD

But this assumes that in English all letters are used with the same frequency, which is not the case. E, A, and R are the three most frequently used letters at rates of 11.1607 percent, 8.4966 percent, and 7.5809 percent, whereas a letter like X only crops up 0.2902 percent of the time. So, Shannon then tweaked the way he drew letters, by forcing 11.1607 percent of his otherwise random finger pointings to produce the letter E, and so on for all the frequencies of letter use in English. His next construction then looked like this:

OCRO HLI RGWR NMIELWIS EU LL NBNESEBYA TH EEI ALHENHTTPA OOBTTVA NAH BRL

The third step was what he called the "second-order approximation," which added what's termed a "digram structure" (also known as a bigram). This is the fact that certain 2-letter combinations are much more likely than others. In English the combination "TK" almost never happens, but "ED" happens often, and "QU" is basically compulsory. To make his life easier, he ditched the random number table for this stage and instead used a book (we don't know what the book was). In this he flipped to a random page, selected a random letter, then flipped to another page and read until he hit that letter and took it and the one after it. He then flipped to another page and looked for that second letter, repeating the process until he got:

ON IE ANTSOUTINYS ARE T INCTORE ST BE S DEAMY ACHIN D ILONASIVE TUCOOWE AT TEASONARE FUSO TIZIN ANDY TOBE SEACE CTISBE.

It's pretty garbled but it's definitely getting more interesting. The words "ON," "ARE," "BE," "AT," "ANDY" (if one allows names), and "TOBE" (a northern African cotton garment) have magically appeared. At the next stage Shannon used the trigram structure for English, obtaining this:

IN NO IST LAT WHEY CRATICT FROURE BIRS GROCID PONDENOME OF DEMONSTURES OF THE REPTAGIN IS REGOACTIONA OF CRE

Obviously, one could keep going, using the measured frequencies of increasingly long letter groupings of 4, 5, 6, and so on up to any arbitrary N-gram. But without a computer to help him, Shannon leapfrogged to randomly picking words from a book, but weighting those choices by their known frequency of occurrence in English, or their probability.

> REPRESENTING AND SPEEDILY IS AN GOOD APT OR COME CAN DIFFERENT NATURAL HERE HE THE A IN CAME THE TO OF TO EXPERT GRAY COME TO FURNISHES THE LINE MESSAGE HAD BE THESE

And finally, he used the second-order *word* approximation, or word digram. Picking a word at random in the book, then flicking ahead until he found another instance of it and using the word that followed immediately afterward:

> THE HEAD AND IN FRONTAL ATTACK ON AN ENGLISH WRITER THAT THE CHARACTER OF THIS POINT IS THEREFORE ANOTHER METHOD FOR THE LETTERS THAT THE TIME OF WHO EVER TOLD THE PROBLEM FOR AN UNEXPECTED

All of which sounds like a crude machine learning attempt at writing literature (which may not be coincidental because, for the aficionados, this is a type of discrete Markov process, something that crops up in modern machine learning). But what seems like Shannon having a bit of peculiar fun is actually deeply instructive.

First, it has to be emphasized that he could have gotten to this final sentence solely by adding ever more levels of measured letter association frequency—more of the N-grams. His switching to using whole words was simply a shortcut because, unlike us, he didn't have a fast, multipurpose computer in his hand or on his desk. Critically, that final, almost-normal-sounding sentence is entirely the result of random choices that are weighted according to nothing more than the known statistics of the English language. Which is pretty astonishing. Our precious letters and words can be corralled into a mathematical enclosure, the sum total of human literature and self-expression squeezed into rules of probability.

Second, it's easy to see the progression from *meaningless* to *meaningful* between the very first random jumble and the last sentence. But in strict informational terms that progression is also about going from *more* information to *less* information, because the final sentence is a result of applying all the predetermined weightings of the English language. It is, in effect, like tossing a bunch of seriously biased coins.

The curious (but perhaps positive) news is that the final, more meaningful sentence has more redundancy. If we step back for a moment to think about the compressibility of data, this last sentence is much more compressible than the first one. That first sentence has more surprise because you'd be hard pushed to predict what letter is coming next. If Shannon did his random pointing properly you should only have a 1 in 27 chance of predicting each character. By contrast, the last sentence—which is most interesting to us as humans—is less full of surprise, and is therefore easier to transmit or to store, if you know the rules of English. In this sense it is just like the earlier example of a checkerboard. That grid of light and dark squares has considerable meaning, and is quite compressible.

For the icing on Shannon's cake, his theoretical analysis yielded a formula for a specific numerical quantity that can be calculated for any string of transmitted data, or any dataset at all. The full details are not necessary to grasp the implications (the endnotes contain more of the gory details). In short, the formula lets us measure the average minimum storage (as bits) required for the information in, for instance, a piece of text *given* what the text contains. Looking at that text tells us how frequently a particular letter is used—the letter's probability of appearing at any character space. If a letter like "A" has an 80 percent probability of appearing at any place, then most of the text is filled with "AAAA"s. That simplifies the text and reduces how many bits are needed to denote all those boring, repeated letters.

A couple more examples can help us see how this works. Suppose I feed the entire text of Shakespeare's *Hamlet* into a code to compute the result of Shannon's formula. The answer to three decimal places is: 4.468. This number is expressed in units of "bits per symbol," which tells us how many bits (binary 1s or 0s) are needed on average to uniquely code each symbol or letter in the entirety of this text. Letters that are much more common (like E in English) require fewer bits. Now what on earth does that really tell us?

We can do more to see the pattern. The first couple of pages of James Joyce's *Ulysses* yield a result of 4.366. John F. Kennedy's "We choose to go to the Moon" speech, given at Rice University in 1962, manages a figure of 4.241. The lyrics and chorus to Aerosmith's 1975 song "Walk This Way" manages 4.225, and Dr. Seuss's wonderful and hugely repetitive *Green Eggs and Ham* clocks in with 4.181. In strict terms of informational surprise, James Joyce beats out Dr. Seuss's addictive, but highly repetitive rhymes. That means that James Joyce is less redundant, and contains more absolute information than Dr. Seuss.

Another way to think about this is to imagine a piece of a crossword puzzle, a row of boxes into each of which you can place a single letter. If you were naïve about how English worked you would say that each box could contain one of 26 symbols with equal probability (a 1-in-26 chance draw from the alphabet). But in reality, as the story of Shannon's original thought experiment shows, different symbols—letters of the alphabet—have different probabilities of being used. So, we should at least assign those unique probabilities to each symbol. The probability of a specific symbol in each box also depends on what's in the boxes before it. If I see the letters "ELEPHAN" and the last box is empty I'm pretty certain that the letter that should go there is "T." That letter "T" may finish the word, but it's also very unsurprising—it doesn't add much information.

Shannon's formula lets us incorporate as much, or as little, knowledge about those symbols' probabilistic relationships as we want. It is a way of evaluating, in a pure, mathematical way, the information content of a crossword answer, a sentence, a book, or the entirety of the human dataome. All at a level (an N-gram, if you will) of your choosing.

And this is where we reconnect with the idea of compression that we talked about with checkerboards and the *Mona Lisa*. The formula is a mathematical measure of how much surprise there is and how many bits are needed to encode some information. Consequently, it also yields a universal measure, or metric, of *compressibility*.

## Entropy, Again

Shannon actually vacillated a lot on what to call his deceptively simple formula. He wondered if it was itself best labeled "information" (in a mathematical sense of how many distinguishable states there are in a representation scheme, such as the ten digits for a decimal system). Or was it better to call it "uncertainty," in the sense of how knowable a value could ever be?

I know, it's a miracle that scientists ever get asked to parties. But there was good reason for this kind of obsession with detail. That's because the formula had an uncanny resemblance to another formula describing a deep physical property of the world. A property we've come across before.

The most widely told anecdote about how Shannon finally found a name for his formula relates the tale of how sometime in 1940 he encountered the mathematician John von Neumann (whom we will also meet again) at the Institute for Advanced Study in Princeton. After pondering Shannon's conundrum for a while von Neumann told him that he should use the term "entropy" because Shannon's mathematical function was effectively the one that already appeared in statistical mechanics to describe entropy in physics. And since most people didn't understand what entropy really was, he would always have the advantage in an argument. In the end Shannon did call his formula "entropy," and expressed it as a measure of uncertainty. Today we simply call it Shannon's entropy.

Whether or not this is exactly what happened, there is indeed a profound connection between thermodynamic entropy and the entropy that is associated with information. It's a connection that's going to appear for us again and again, and it also marks a turning point in how we think about the dataome and the world in general.

The atomic nature of matter, captured by the ideas of statistical mechanics, led to a picture of entropy in terms of the number of microstates in any system, the number of possible arrangements of atoms or molecules. The larger the entropy the more microstates and the greater our *uncertainty* about which specific microstate a system might be in at any instant. That's just like the *surprise* of Shannon's entropy. It is in that probabilistic statement where entropy and information merge. More entropy means more information in the sense used by Shannon.

But all of this may still sound confusing. After all, a while back we saw that entropy is both a measure of the limits of what energy can accomplish in a system and a measure of disorder in a system. Increased information doesn't seem particularly compatible with either interpretation. Indeed, if you poke around in books and articles you will be lucky to find a single universally agreed-upon explanation either for what entropy really, truly means or for its connection to information (and this is decades after von Neumann's scathing assessment).

My best attempt at connecting these dots goes as follows. What we saw with statistical mechanics is that entropy is related to the improbability that your dirty laundry will clean itself or that your cup of cold coffee will suddenly become hot. Both could, in principle, happen, but they are so utterly unlikely that we will never see them occur.

But the simpler a system is at a microscopic level, the less improbable those kinds of reversions are. A single atom can be moved from here to there and back again; there are comparatively few possible microstates, to use the language of statistical mechanics. While work has to be done by the external world to make that happen, the atom itself is perfectly OK being tidied up, and the changes are simple and need not be improbable.

Suppose, though, that we had to write a set of instructions for how to clean your laundry by moving every relevant atom back to where it had been. Or a set of instructions for making your drink hot again by moving every atom back to the state it was first in and reabsorbing every lost infrared photon. Those instructions would be mammoth in scale! Whereas the instructions for moving a single atom back to its original position are, in principle, pretty trivial. Furthermore, every time you do something in the world, like boil more water for coffee or toss another sock into the pile, you are increasing the length of those instructions.

You can think about Shannon's entropy as a way to measure the size of those instructions, and therefore a measure of the thermodynamic conditions they describe. Informational entropy and physical entropy are two inextricably linked sides to the same story.

In essence, what we encountered in statistical mechanics—the idea of matter occupying microstates and macrostates, and our imaginary mail-sorting room and its many pigeonholes with a whirling mail dervish arranging and rearranging envelopes—can be described in the same way that a sequence of letters in a book can be described. Each possible book is a microstate occupied by the symbols of an alphabet. Describing the properties of atoms in a gas or letters in a language seems to converge onto the same tools. Which, if you think about it, is pretty crazy.

There are many more layers of sophistication that can be added. What if, in the above analogy, the instructions have contingencies and dependencies? By moving an atom or an envelope I may alter the instructions for the next step by reducing the choices for the next move—just like the example of ELEPHANT. In many languages there is also a difference between fine-grained information and coarse-grained information, with one influencing the other. An example would be to say that I have a basket of citrus fruit (coarse-grained), versus saying that I have a basket of three oranges and five lemons (fine-grained). As with many things, Shannon's entropy is just the beginning of the story.

## The Meaning of Meaning

Critically, Shannon's entropy metric does not tell us anything about *meaning* beyond the probabilistic usage of symbols in language or data. It is, as I've said, a purposefully austere mathematical evaluation of information. Entropy might discriminate between Dr. Seuss and James Joyce, but it cannot tell us what the impact of either's writings really is for any of us. Meaning is key, and as I've suggested, when humans talk about meaning we're usually referring to things that speak to our personal experiences, our aesthetic senses, and our intellects, or to our emotional inner lives and our ability to enhance our sense of satisfaction and happiness.

But there's quite a gap between these very subjective statements and Shannon's stark mathematical measurement of entropy and surprise. Is there something in between? A measure of meaning that is a little more than probabilistic surprise, but not as subjective as a zoomorphic cloud or a poetic phrase?

One possibility emerges from stepping back and looking at how information is *used* in the world. Such an approach implicitly involves the idea of living things or algorithms that cogitate or compute using information. The physicist Carlo Rovelli, writing in 2016, beautifully reviewed and summarized the outlines of just such a way to think about information, one deeply connected to the principles of Darwinian evolution.

Imagine, Rovelli suggests, a simple organism that is looking for food. Perhaps there is only food sitting to the right-hand side of where the organism is. If the organism has that information and can act on it to move to the right, then its *survival* is surely improved. That information about food has a direct meaning that is not the same as for information about the color of the sky or the time of day—at least in these circumstances. It is information that is correlated with the external world and with what the organism needs. In the long term a randomly moving organism will not succeed as much as an organism that moves according to where the food is (assuming that its food is rare).

Here, meaningful information is information that influences the processes of natural selection. Exactly how this measure of meaning changes as we look at increasingly complex systems and hierarchies will, as I'll show, turn out to be centrally important to understanding the very nature of life itself.

For now, it's worth pointing out just the beginnings of that deep story. When I say that an organism "has information," what I mean is that it encodes something about the external world in itself. It has to, because how else can it act on that information? To use a simple analogy: when you make a to-do list you are taking information about the external world and manifesting it as ink on paper or electrons in a computer. You're remapping part of the world into a new substrate. You also do this in your brain whenever you create a memory or react to something.

In this sense, meaningful information for survival becomes correlated between two places: the world and an organism. Remarkably, Claude Shannon already tackled the mathematics needed to probe this. The idea is called *mutual information*, and it measures the amount of information one entity can give about another. It doesn't explain how or why two things—whether mathematical variables or organisms—are related, but it does quantify how knowledge of one thing reduces uncertainty about another thing. That in turn brings us back to one of the fundamental challenges that motivated Shannon's efforts in the first place.

## Talking

Let's go back to that attempted conversation with Alice across a busy street, and the exchange of information. The biggest obstacle is noise, the hubbub that forces you to raise your voice, and what caused those early telegraph lines to burn out as the voltage was cranked up.

The best fix for this, largely due to Shannon, is almost the reverse of distilling the information content of a signal, of evaluating its compressibility, because it involves adding redundancy back into the transmitted data. But not just any redundancy: careful, mathematically precise additions to the way in which data is encoded which enhance the probability that a piece of a transmission will be correctly decoded even if noise corrupts it to some degree.

A simple example helps explain this. Imagine an alphabet that uses four letters, A, G, C, T. The laziest, most basic way to encode these letters in the language of bits and binary numbers is as:

A=00
G=01
C=10
T=11

But if any digit gets corrupted in transmission, changing from 0 to 1 or 1 to 0, it's impossible to know what the truth is. An "A" can change into a "C" with the simple flip of a 0 to a 1 in the first position.

Shannon's insight was that we can, with care, make the encoding more robust, at the cost of some extra data. The following encoding is much more resistant to noise:

A=00000
G=00111
C=11100
T=11011

In this case even if A suffers an error in transmission, becoming say 00100, it still doesn't look like any of the other letter codes, and it's most likely to have originally been an A. It takes 3 erroneous bit changes (0 to 1 or vice versa) in the digits to turn any letter into another.

The practical applications are immense. Even though Shannon couldn't show exactly what the best kind of encoding of information would be in any particular case, he did show that there must always be a best encoding, we just have to be smart enough to figure it out. In the context of our own human languages this is also a provocative idea. It raises the question of whether some languages might be more intrinsically robust than others. Intriguingly, as we'll see, all humans may actually converge on very similar language efficiencies.

The reason I use the letters A, G, C, T in the above example is to also point out that biology seems to already know about these tricks. In this case the four nucleotides of DNA—adenine, guanine, cytosine, and thymine—are the letters themselves. They in turn make the three-letter "words" that correspond to just 20 common amino acids used by all of biology (plus a couple of rarer ones and so-called "stop" codons). But 4 letters combined into 3-letter words could in principle correspond to 64 unique words or amino acids, which would've been a much richer language.

Instead, nature has converged onto a redundancy in its root coding. Some amino acids like leucine correspond to six differently spelled three-letter "words," others to four, two, and a lone one-word coding. Our first guess as to why evolution has hit on this scheme is that it provides tolerance to the noise of "point mutations," in which a single nucleotide gets corrupted (an A becomes a G, for example). Differently coded words may also change the rates at which parts of DNA are transcribed, and alter the ultimate shape of proteins made in cells. All of which further hints that Shannon's insights opened a window to a profound informational foundation for how the universe works.

This also brings us all the way back to the utility, or health, of data. Compressibility and Shannon's entropy provide a baseline measure of the intrinsic information content of data. But the *health* of that data must also relate to how robust it is; how well encoded it is to withstand noise and corruption. And when an organism's data contains mutual, survival related information about the organism and its environment, the meaning held in that data must also reflect on that data's health.

All of the intricate and tricky ideas in information theory lead to a fascinating proposition for the human dataome as the bridge between us and the informational content of the physical world. Instead of correlating ourselves with all of that external content to aid survival, like a bacterium, we utilize a third party to take on much of the load. As much as our biological selves have meaningful mutual information with our environment, the dataome is increasingly where that burden lies. But if something goes wrong with the dataome, or if our channels of communication with it get too noisy, that could be a big problem.

## Dataome Entropy

Today information entropy has become a core piece of a much more extensive mathematical toolkit that includes things like Shannon's "source coding theorem." That theorem tells us just what the theoretical optimum level of data compression can ever be, according to the coding (the alphabet) used. Machine-learning scientists further trick out Shannon's entropy and its associated ideas with all manner of inventive twists. They use it to quantify the informational trade-off between compressing or sampling data and accurately capturing that data's essentials for training deep-learning systems to make predictions. They connect artificial neural networks to the rules of game theory, in which the outcomes for competing entities are contingent on what the other player does. So-called genetic algorithms seek optimal solutions to complicated problems by simulating survival-of-the-fittest trials. All these approaches pay lip service to the ideas of mutual information and self-information (the surprise of the value of a variable within a system).

Researchers get excited by these tools because they can create new value in otherwise impenetrable data. We can look for complex features in vast troves of medical and genomic data, seeking the best self-information—the most surprise. Or we can use Shannon's entropy for evaluating the volatility and surprise of financial markets, and even for pricing assets. Entropy-gauged information is a starting point for evaluating the health of data before worrying about context and meaning. In that sense it is a triage measurement that everything else follows on from.

In 2011 the researchers Martin Hilbert and Priscila López published an absorbing and eye-opening study in the journal *Science* with the title "The World's Technological Capacity to Store, Communicate, and Compute Information." The way the authors tackle this subject involves some clever footwork and, of course, Shannon's entropy.

Across the timespan from 1986 to 2007 the authors make detailed estimates of the global hardware capacity for data storage—in books, photos, hard drives, memory cards, and everything else. They also estimate the global capacity to broadcast and communicate information, whether by newspaper and radio and television, or through telephones and the internet. But this necessarily mingles decidedly old-school, analog technologies with digital ones. It also mixes the history of how we've optimized, or compressed, our data. For instance, in 1993 a major algorithm for compressing video was called Cinepak, but it was at least three times less efficient at compression than the more recent MPEG-4 algorithm that you'll likely find you're watching today.

To get around this the authors point out that what really matters is not the amount of data that can be stored or moved around, but the amount of information. And information can all be brought to a level playing field of sorts by determining its Shannon entropy. Doing so in effect tells us whether or not the whole of the human data repository actually just consists of the same amusing video duplicated a trillion times. To approximate the task of computing the entropy of so many disparate sources, Hilbert and López simply ask what the optimal compression technology used in 2007 would produce across all the years they studied.

The answers are fascinating, albeit a little confounding. In 1986 the optimally compressed information storage capacity per capita across the world averaged out at 539 MB. By 2007 it had leaped to 44,716 MB. In terms of a global total that means that in the year 2007 humanity had the ability to store $2.4 \times 10^{21}$ bits of non-redundant information, or about 300 exabytes (where an exabyte is a trillion megabytes), in all of its technological devices.

The post-2007 growth rate has remained pretty steady. By 2016 data center storage alone was topping out at around 800 exabytes. In 2020 it crested over 2,000 exabytes. Some estimates suggest that the total number of words spoken by all humans that have ever lived, across all their lifetimes, comes to around 5 exabytes worth of data. In just one year we can now exceed that by a factor of 400.

Altogether Hilbert and López found that the world's capacity to store and communicate information (measured to a common standard by Shannon's methods) had experienced a roughly 20 percent to 30 percent annual per capita growth rate in the period they studied. That rate appears to have been maintained in subsequent years. Similarly, the world's per capita technological capacity to transmit and receive information—to telecommunicate, including across the internet—also grew at about 28 percent annually.

But the thing that has accelerated to an even more extraordinary degree is the per capita technological capacity to compute: to chug electronically through calculations, comparisons, and algorithms. Whether via human-mediated computation or fully automated processing, this compute capacity grew between 58 percent and 83 percent annually over the timeline of their study, depending on the precise gauge used to estimate it.

In other words, as much as the human dataome is seeing remarkable growth in its non-redundant size and mobility, even more remarkable is the growth in the ability to compute information—to process and interrogate data. On the face of it this suggests considerable healthiness to the dataome. Yes, it's growing bigger and bigger, but our primary tools for interacting with it are gaining capacity faster.

However, this could also reflect a chicken-and-egg puzzle. Growing data storage capacity and data size might be the primary driver of better computing capacity, but it could also work the other way. The more computational oomph we have at our disposal, the more we can see doing with it, and so the more we look to gathering and storing data. Much as with the tale of Big Data, once we think we have a useful tool, we tend to apply it as much as we can.

Or, all of these trends could be accurate but still not quite capture the bigger picture. We saw in an earlier chapter how the rate of growth of printed material has, across the past few centuries, easily outstripped the growth of its primary consumer—namely the global human population. Superficially, this is the polar opposite of what's happening with the growth of computing power (consuming data) compared with data storage and transmission.

It could simply be that the evolution of data volume and computing capacity have not yet reached an equilibrium state. Many hard-to-foresee human decisions drive what we develop and invest in. The growth of computation, from massive servers to smartphones, has a potent but complicated connection to commerce and economics, as well as to culture. It's also related to the geopolitical stability of the world during recent times. None of these things are fixed points. If the world is in perpetual disequilibrium there is no reason to expect alignment between the growth of computing and the growth of data storage and movement capacity.

But another, more worrisome possibility is that the *meaning* of information held in our data is actually falling behind. As a result, we are relying on faster growth in our computational capacity to try to dig out useful stuff, whether we realize it or not.

## This Statement Is False

On the face of things your information doesn't count for much if it's inaccurate. There is no obvious evolutionary benefit to chasing false clues to food or false guides to predator evasion. But humans walk a wavy line when it comes to the truth. We're finely tuned to picking up falsehoods, but we're awfully good at lying, quite prone to being deceived, and perfectly capable of cognitive bias. Much of that doesn't even happen face to face but is instead mediated and enabled by the dataome.

For a very long time we've published opinion and prejudice as fact in papyrus rolls or pamphlets, and in books and newspapers. Today we send digital messages pretending to be money-laden princes or distraught friends in order to exploit complete strangers. Advertisements bend the truth in glossy images and videos. Nations mount disinformation campaigns through social media and fake identities. And confused individuals and organizations propagate anti-scientific, anti-rational nonsense on everything from diets to vaccination, and whether or not the Earth is flat (it isn't, by the way).

It's reasonable to wonder at what point all of this deception might become unstoppably detrimental to a species, and whether the dataome creates a unique vulnerability that evolution has yet to prune away. This is a fundamental question. Data falsehoods are nothing new for living systems. The ability to manipulate information and deceive has been an element of life on Earth for far longer than modern humans have existed. As a result, the natural world contains a barrage of information that has either been weaponized or declawed, and life's balancing act between accurate and inaccurate information could provide some clues about the dataome and its future.

Bacteria provide one good example. These single-celled organisms have innumerable ways to disguise themselves or misinform other species in their environment—including the environment represented by multicellular life. The coevolution of animals and pathogens has resulted in microbes becoming masters of biological stealth. They can alter the structure of their cell membranes, or release proteins that inhibit or degrade immune factors in a host organism, and they can mimic host chemistry to blend into the background.

*E. coli,* for instance, can chemically engineer their surfaces to switch from a positively electrically charged exterior to a negative one, which selectively repels some of the molecules produced by a host's immune cells. They also secrete a special protein molecule that in turn tweaks an enzyme in the unfortunate host to modify certain immune responses. As a result, *E. coli* are not only better at evading an animal's immune system, they actually keep the host alive for longer, reducing the symptoms of infection, so that the bacteria have a greater chance of spreading on to the next victim. Other strategies are equally insidious. *Neisseria meningitidis* switches off genes that coat it in a polysaccharide capsule in order to enter the bloodstream, then switches them back on to cloak itself as if part of the host system.

Camouflaging and spoofing, or simply confusing other organisms to aid infiltration and survival, extends all the way up the chain of increasingly complex life. Certain succulent plants in Namibia grow to look astonishingly like pebbles, a trait thought to help avoid being eaten by foraging creatures. There are insects that camouflage themselves as twigs and leaves, or mimic other, more fearsome insects. Certain species of fish have markings that look like the eyes of a larger creature. There are various species of snakes whose tails can mimic their heads in appearance and movement, distracting would-be predators from their more vital parts. There is even the menacingly named spider-tailed horned viper of western Iran, whose fronded reptilian tail is a convincing enough spider simulacrum that it can lure hungry and unsuspecting birds to their doom.

These kinds of false information are all in aid of predation or predation avoidance. They are also excellent examples of mutual information, where species encode sophisticated information about the world around them in support of survival. Humans are no different, but we do take some of this to a new extreme of cognitive sophistication. The ancient text of *The Art of War*, attributed to Sun Tzu, a general and strategist around 500 BC in China, advocates deception as a central strategy: "All warfare is based on deception. Hence, when able to attack, we must seem unable; when using our forces, we must appear inactive; when we are near, we must make the enemy believe we are far away; when far away, we must make him believe we are near" (chapter 1, verses 18–19).

Perhaps reassuringly, outright deception is by no means a default property in living systems. Indeed, in sexually reproducing species, the phenomenon of honest signaling seems to play a central role in what is called sexual selection—the critical ways that advantageous reproductive choices are made. The only cautionary note is in the word "seems." As with many areas of evolutionary theory, there remains a great deal of uncertainty about the details.

A commonly cited example of honest signaling is that of the birds of paradise. There are some forty species in this avian family, mostly in Indonesia, Papua New Guinea, and Australia. They are often strongly sexually dimorphic, with the males having elaborate, brilliantly colorful plumage and a variety of equally elaborate approaches to displaying themselves in the quest for a female mate. Often this involves outrageous dancing performances, at a level of intricacy and species specificity that makes human night clubs look positively pedestrian. By comparison the females tend to be visually muted, without elaborate plumage or behaviors.

In the general framework of Darwinian selection and evolution, these signaling traits surely exist because they confer some kind of advantage, or else they would be pruned away. The received wisdom in evolutionary biology is that the glorious plumage and snazzy dance moves of the male birds of paradise are *honest* traits. That is to say that if bird #1 (call him Bob) has larger, more colorful feathers and a better dance than bird #2 (call him Ishmael) this really does accurately reflect Bob being a healthier, more genetically fit specimen, better able to provide resources to ensure the survival of offspring, and with fewer detrimental gene variants. Therefore, when the female subject of Bob's and Ishmael's attentions evaluates the two of them, she can make a choice that genuinely increases the odds of having more and fitter offspring carrying some of her genes into the future.

The detailed mechanics of this are bundled into a subfield of evolutionary biology called signaling theory. We can mathematically model the properties and outcomes of signaling strategies to try to understand how these display patterns emerge. It's a fascinating area of study but far from stable itself, with a variety of competing ideas out there. One important area of debate involves the cost, or burden, of honest traits. While Bob may be the fitter mate, his plumage is also more biologically expensive than Ishmael's drabber outfit. Bob will need to gobble up more food as a result, and has a harder time flying away from predators because of his beautiful but cumbersome tail feathers.

This additional burden on Bob only seems to makes sense if, in the end, there is a statistical advantage for his lineage, with the costs outweighed by the benefits. If, as an alternative, male birds of paradise could produce fabulous feathers and dances with little effort, there would be no constraint on their spoofing or cheating. It would be a feathery free-for-all.

This latter scenario would also mean less-fit offspring would be filling up the bird population, and so the actual viability of the species would decline. This observation has led to an idea known as the handicap principle. It seems counterintuitive, but greater biological fitness has to (in general) *also* be signaled by greater handicaps. In very human terms it's a kind of conspicuous consumption. A good example is the peacock, whose tail feathers can be utterly absurd from the point of view of mobility. Yet this can still be considered a genuine, honest marker of genetic health because only the most robust males can survive with such an impediment. The principle doesn't just apply to appearances. The *internal* biochemistry of organisms also relates to risk-taking and "showing off." For example, while increased testosterone can increase muscle mass and aggression it can also suppress an immune system.

There are plenty of complicated gray areas too. A potential mate might be influenced by a combination of factors beyond traits of apparent health and fitness. If the odds of interacting with a good potential partner are slim (in sparse populations, or if the desired trait is rare), there is a search cost. How long do you wait for the right one to come along? The parallel to literal search costs in the dataome is pretty obvious. Information that is easy to search for is not necessarily the best information, but how much effort do you expend looking more deeply?

That search cost, in turn, relates to the question of just how much honesty there really needs to be. Is pure honesty really always the better strategy? Nature, as always, seems to be running a gazillion other kinds of experiments in parallel, some of which speak to this question. A striking example of a type of dishonest signaling is seen in zebra finches—found mostly in Australia and parts of Indonesia. These charming little birds are socially monogamous, forming long-term, but not necessarily permanent male-female breeding pairs. The male zebra finch uses song as a part of its signaling, and song is costly. Singing takes physical energy, cuts into the bird's foraging time, and exposes the singer to predators.

A study carried out in 2013 suggests that during the first minutes of courtship, males of dubious fitness can nonetheless muster themselves to produce songs that females will react to as honest traits—being duped into spending more time together. However, the male doesn't get to put its little feet up in the relationship. The female is much better at discerning real quality in the male over longer periods of a few days. Lengthier interaction allows them to mostly, but not always, sort the wheat from the chaff. Because it's exhausting, the less robust males simply can't keep up to the level of song they use in the initial encounters. The result is a complex game of chance, sometimes working out to the scruffier males' advantage, but more often benefiting stubborn females or truly classy males.

This is a vivid example of the kind of fine-tuning necessary for a species to continue to exist. There are exquisitely delicate balances at play between the various subtle elements of zebra finch behaviors. We might see the deceptive finches as a flaw, but that may not be true. It is possible that there is a role for inaccurate information, as long as, on average, it is held in check over time.

Edging a little closer to the kind of cognitive sophistication that's at play in the human dataome are proposals for the idea of aesthetic selection in the natural world; a concept that Darwin noodled with back in the 1870s. More recently, in 2017, the ornithologist and evolutionary biologist Richard Prum posited that potential mates might have preferences for certain traits, such as feather colors or other appearances, that are entirely disconnected from biological fitness. A trait is simply appealing, for reasons too random or complex to really decode, just like our own age-old quandaries over quantifying the notion of beauty.

Not only might those aesthetic preferences result in the generational selection of certain traits, those preferences might be inherited by offspring, a kind of runaway coevolution. If the aesthetic preference is strong enough, and heritable, it could totally undermine the fitness part of natural selection. Not surprisingly, this is all pretty controversial, in part because we simply don't know how something like aesthetic taste can be encoded genetically. But for humans it seems perfectly conceivable that our dataome plays a role in reinforcing aesthetic or value-based preference between generations. That could be beneficial or detrimental. It doesn't take much imagination to think of instances such as climate-change denialism or anti-vaccination beliefs that are appealing to some people—and even influence their partner choices—but hugely dangerous to the species. In which case problems with the overall validity and accessibility of information in the dataome could very well be undermining critical aspects of our species' evolutionary fitness.

Honesty and dishonesty, in the form of accurate and inaccurate information, clearly matter at multiple levels in the biological world. But for humans the informational environment includes the dataome, and it impacts our decisions and our preferences, as individuals and as groups. Throughout human history we've been filling the dataome with information of widely varying quality—both in the quantitative sense of Shannon's entropy and in terms of absolute accuracy and the harder-to-quantify property of meaningfulness. Just like the little zebra finches and other species, we are participants in a dynamically evolving system. The apparent imbalance at present between growth in data storage and growth in computation may be perfectly fine if that system balances out in the longer-term, statistical sense.

Equally, there does seem to be something genuinely new going on, as the past century of our interaction with the dataome has increasingly required machine intermediaries and translators. The quintillions of daily bytes cast into small electrical charges or magnetic zones represent a vastly different reality than anything biology has equipped us to deal with. As a result, the opportunity for deception and obscuration of information is extremely significant. The question is whether we're populating our dataome with tools and agents that act for the greater good of humans or for something else. If there is no fine-tuned statistical balance in place in the long run we could be in deep trouble.

## Adversarial Truth

As I write this text there are machine-learning algorithms out in the world that can competently translate languages or generate new music in the style of famous composers. Some algorithms can even construct text that mimics the writing style of real people, or accurately mimic specific human voices. Others take still or video images of humans (real or imagined) and generate believable "deepfake" versions of them expressing any view and any agenda.

Many of these capabilities come from a machine learning approach called Generative Adversarial Networks (GANS). The essence of the approach is simple: two machine learning algorithms are set up in competition. One generates data and the other tries to classify that data. A machine learning system may be configured to learn how to paint pictures like Vincent van Gogh, using examples of his work. Another system may be set up to learn how to discriminate between van Gogh's paintings and any other images. The two machines are then pitted against each other. The painter machine produces its best effort at a van Gogh lookalike, the critic machine examines that and returns an assessment of how likely it is to be a real van Gogh. The painter then tries again, aiming to improve its score. The process repeats until painter and critic are doing really well at their tasks.

Unlike humans, the machine learning systems can repeat this process millions of times, and track the salient details of what they did each time. It is nothing short of miraculous to see this play out. It is also horrifying to see how well a set of mathematical rules and digital switches can end up mimicking qualities that we hold so dear to ourselves as *Homo sapiens.*

The behaviors of GANS are eerily similar to the kind of evolutionary tugs of war I've outlined above. In those biological cases there is an endless back and forth between the honesty or dishonesty of signaling and the responses of those being signaled. But in biology an equilibrium of sorts is established at any given time, or else a species would cease to exist. For GANS and our deployment of them in the dataome, it is far from clear that there is any kind of equilibrium—at least not yet.

While GANS are, at present, not "in the wild" as reproducing species, what they generate is increasingly flowing into the human dataome. As I've described, fake news and fake data are hardly novel things for us. But what makes GANS fakes different is that they are invading a realm of human cognitive experience previously more immune to dishonesty, and they are capable of pumping out vast amounts of fake material very quickly.

I'd claim that GANS fakery has the potential to exponentially reduce the intrinsic value of information within the dataome. It's easy to see that if it targets political viewpoints and social control it will cause immediate real-world damage (potentially devastating damage in both social and physical forms.) But even if it only becomes a tool in our daily technological repertory, like photo filters, auto-tuning, and animated GIF generation, it will have an equally profound impact. The only way to fully counter GANS-generated material is with other GANS, because that's how the best possible discriminators of fakery (the critic machines) can be developed. For humans to continue to benefit from the internet, from Big Data, from our connectivity and access to the dataome, we will need to know what is real and what is not.

From a geopolitical point of view, controlling or manipulating what your country's population sees, or what another country's population sees, is a powerful and immensely seductive tool. It's also not clear that any of us can do much to limit the use of systems like GANS. They are eminently weaponizable. Thus, nation-state defense systems are likely to be developed (if they're not already) that employ GANS. The alarmist picture is that the same kind of arms race that is used to produce a single GANS system, and the same kind of arms race that is going on all the time in biology, will spill out into the world at large. In biological systems, if dishonest signaling doesn't make sense a lineage will be pruned out of existence. Is there a pruning mechanism in the dataome?

In a localized sense there certainly could be. If the generation of fake videos becomes too expensive because machine-based discriminators can always spot a spoof, then they'll lose their power. Or if fake videos are provably impossible to verify, some people might lose interest in the medium, producing a kind of herd immunity—although that is likely hugely optimistic.

There are, however, signs that some of the most sophisticated AI systems, so-called deep neural networks (DNNs)—which can excel at image recognition, among other things—are also quite easily fooled. They can be brilliant at a well-defined task like recognizing traffic signs or animals in photos, but they can be astoundingly brittle, too. Faced with totally unexpected, adversarial data, they still produce an answer with confidence, even if it is completely nonsensical. Perhaps then a system like a GANS will prove unable to truly generalize its actions, curtailing its spread.

Even before such pruning points are reached there will likely be fierce competition between fake information and tools like internet search engines. That leads to one predictable outcome, which can be stated with high confidence: the more our dataome becomes inundated with fakery, with machine-driven arms races, the greater the burden on our planetary resources. Anything that reduces the meaning of human information threatens the balance (delicate or otherwise) between us and our future selves in a way no less profound than in biological evolution.

## The Mortal Coil

Data can of course also be lost, erased, and forgotten for good. We seldom give thought to what most ordinary but literate people of Elizabethan England were writing about while Shakespeare lived. There must have been innumerable letters, merchants' notes and records, legal documents, stories, poems, and all manner of materials that were only read or used by a few people, then destroyed by the erosions of time. That's a story that has played out across all human cultures and populations, regardless of our technology. What filled the burgeoning internet in the 1990s is increasingly hard to find as companies have gone bust and services have disappeared. Web pages vanish when an outdated hard drive finally keels over. Media like magnetic tapes and floppy disks become unreadable because of a loss of integrity or a loss of the very machines that were used to read them. The internet pioneer Vint Cerf calls this phenomenon "bit rot." Despite the astonishing growth rate of storage and computation, data is constantly disappearing. Information is pruned away, just as it has been throughout human history.

Where does all of this leave us? In the last chapter I outlined the burden of human data and its contribution to our footprint on Earth. Here I've looked into the root worth of data, the quantifiable value and meaning of information for living systems, and ways in which it can be degraded. The most extraordinary discovery is that information isn't as nebulous a concept as we might have suspected. Shannon's mathematical tools allow us not only to probe and evaluate information, but to make it better and more robust. These same tools, in a wholly unexpected fashion, are fundamentally linked to the rules of matter—the statistical descriptions of atoms and molecules that tell us how stuff behaves in the real, physical world. All signs point toward the human dataome as something far deeper and more intriguing than we might have ever suspected.

But, just as in biological systems, the potential benefits of a dataome are dangling off the other end of a complicated balance. It seems straightforward enough to ask what those benefits are, but the route to an answer covers some tricky terrain. That journey involves thinking about the dataome as a bridge between us and the world, and each other.

# 4

# AN EVER MORE TANGLED BANK

*You think your pain and your heartbreak are unprecedented in the history of the world, but then you read. It was Dostoevsky and Dickens who taught me that the things that tormented me most were the very things that connected me with all the people who were alive, or who ever had been alive.*

—James Baldwin (1963)

Having the ability to carry far more information into our collective future than is encoded in our DNA ought to confer some extraordinary advantages to *Homo sapiens.* Some of these advantages are more obvious than others. The fact that I can write these words on a small, off-the-shelf computer that could outperform the mightiest supercomputer built a mere half-century ago is one indication of the dataome's magic. Or the fact that my body and mind have been tended to over my lifetime by doctors and teachers informed by a vast and dispersed library of medical and academic information accumulated across centuries. I, like billions of others, have been made healthier and more intellectually capable as a result.

But what can seem like a benefit to us in this moment is not necessarily a genuine benefit in a scheme of evolutionary processes that play out across a multitude of timescales. My amazing personal computer may or may not be acting to enhance the propagation of our entire species into the future. My personal privileges of health and education won't be helpful if I end up as a megalomaniacal tyrant who launches a global war. To see beyond any of our day-to-day experiences and explore the grander scope of human evolution—past, present, and future—means taking a look at many more facets of the interplay of information and our lives.

To further complicate matters, there are possible counterexamples from the past where a dataome didn't confer much advantage to a species. As I've mentioned, evidence exists that over 65,000 years ago at least one other branch of hominins, the Neanderthals, were also constructing elements of their own dataome, in their cave paintings, and what seem to be decorative or abstract arts using objects like mollusk shells. The fragmentary nature of those finds leaves room for much more yet to be discovered, of who knows what sophistication. But the Neanderthal species didn't persist, and its dataome has been reduced to these exceedingly rare paintings and cognitive fossils. Luck, timing, and maybe the behavior of others of the *Homo* genus appear to have relegated these seemingly sentient beings to nothing more than a modest genetic contribution to the genome of many of us living today. If having any kind of external data is such a big help, it certainly doesn't seem to have tipped the balance for them.

One way to explore what benefits a dataome brings is through the connections that external information and knowledge have to our concepts of intelligence and cognitive ability. This is not an easy place to go. There were already hints of the quagmire when Darwin published his *On the Origin of Species* in 1859. Famously, he ends the book talking about a "tangled bank" (meaning a hedgerow) as a metaphor for life on earth. It's been quoted so many times that I feel a little lazy adding it here, but I think it matters in this case:

> It is interesting to contemplate a tangled bank, clothed with many plants of many kinds, with birds singing on the bushes, with various insects flitting about, and with worms crawling through the damp earth, and to reflect that these elaborately constructed forms, so different from each other, and dependent upon each other in so complex a manner, have all been produced by laws acting around us. . . . There is grandeur in this view of life, with its several powers, having been originally breathed by the Creator into a few forms or into one; and that, whilst this planet has gone cycling on according to the fixed law of gravity, from so simple a beginning endless forms most beautiful and most wonderful have been, and are being, evolved.

It's a final flourish of poetic imagery that is both expansive in scope and very literally down-to-earth in its reference to humble shrubs, insects, and worms meshed together in the grand experiment of life. All of Darwin's choice phrases are in service of his great insights on the central qualities of biological evolution, including the sheer diversity of living systems. It's hard to think of a better ending. But just before that concluding paragraph Darwin also writes this sentence:

> And as natural selection works solely by and for the good of each being, all corporeal and mental endowments will tend to progress towards perfection.

Compared with the even-handedness of his final summary this is a pretty jolting statement. The idea of perfection in biology is fraught, even if Darwin's main point was about the *progress* toward something ultimately unobtainable, and even unknowable. At its very worst the notion of evolutionary perfection has turned into arguments supporting eugenics and destructive imperialism.

When I've quizzed modern biologists about evolutionary progression, they're much more likely to say that natural selection only evolves species toward being "good enough." If you're okay at getting your genes into future generations then your work is basically done at that moment. The environment (meaning absolutely everything that influences your propagation odds) will always be changing at some level, so a species is never finished evolving, and what might seem close to perfection in one moment will be an excuse for the neighbors to one-up you in the next moment.

Or to put this another way: characteristics that look like perfect solutions to survival can result in inflexibility, and are often too costly to attain in the first place. That's true biologically as well as in human society. When Darwin throws "mental endowments" into the mix, our modern alarm bells also go off. That's not just because of political correctness; there are good reasons to suspect that a perfection of intelligence (itself ill-defined), or steady progress toward it, is not a quality that is guaranteed to confer exceptional advantages. That statement is more than just an oblique poke at humanity's tendency toward self-destruction. It's based on biology.

Take the rubbery little sea squirt: a rooted, filter-feeding, sack-like creature, centered around a tubular siphoning system. Not only is it not terribly high on our list of intelligent organisms (by any reasonable measure, not just as a prejudiced comment about invertebrates), it is known to actually downgrade its smarts during its life. Baby sea squirts—the millimeter-long larvae—start out with a cluster of some 177 neurons. That simple brain, or cerebral ganglion, helps the young sea squirt get around. At this stage in its life cycle it actually resembles a tadpole rather than its adult form of marine couch potato.

That its juvenile self looks like the young of a distinctly vertebrate animal is odd enough, but what happens during its transformation to an adult is odder still. At some point, after swimming about, the baby sea squirt anchors its head onto a suitable underwater rock and undergoes a remarkable set of changes. It appears to reabsorb its tail, its sensory organs, its nervous system, and most, if not all, of its teeny brain.

From a practical, mechanical point of view this is a perfectly reasonable thing to do. Neurons are energy hungry, and who needs to be intelligent when you're a water-filtering bottom-feeder? This strategy clearly works very well. Sea squirts have a lineage that connects to us vertebrates via a last common ancestor some 500 million years ago. In most respects they are as successful a complex animal as we are, if not more so.

Sea sponges are another extreme example. These wonderful animals have neither brains nor nerve cells. But there is genetic evidence suggesting that in the distant past their ancestors did indeed contain neurons. If this is correct (and it is somewhat controversial) then over time they evolved away from that form, driven by the pressures of attaining environmental fitness. Having brains, even simple ones, just wasn't helpful.

Instead of being a coveted sign of ever-increasing perfection, as Darwin wrote, brains and intelligence may only be there when absolutely needed, and dumped when not. Brains are certainly expensive. In a modern human a disproportionate amount of energy goes into running our neural systems, as we'll see.

## Brain Drain

There have been innumerable efforts to gauge our human specialness among species in terms of cognitive abilities and our brain's physiological properties. One rudimentary approach is in the study of brain size versus body size across the animal kingdom, and it's worth looking at this briefly.

If you were a bat, your average body mass would be around 20 grams and your brain would have a mass of about 1 gram, yielding a brain-to-body-mass ratio of 1 to 20. By comparison, a chimpanzee has a body mass of around 90 kilograms and a brain mass of around 0.5 kilograms, yielding a disappointing ratio of 1 to 180. That's ten times lower than for a bat even though the chimp's brain is 500 times bigger. Humans and mice actually have very similar mass ratios of around 1 to 40.

These results suggest that the brain-body mass ratio might not be a terribly helpful gauge of how smart a species is. Bigger animals often need more neurons simply to handle the coordination and sensory input of a bigger body. The gross mass of a brain also tells us next to nothing about the highly specialized nature of its parts, which probably plays a much larger role in determining cognitive ability.

In fact the animals with the absolutely largest brain-to-body mass ratio are ants. While it's true that colonies of ants exhibit powerful, computational problem-solving abilities, the cognitive powers of individuals are at best highly specialized, niche talents. At the other extreme, the smallest known brain-to-body mass ratio for vertebrates is that of the unfortunately named bony-eared assfish. These eel-like fish live at the extreme depths of tropical oceans, in conditions of low oxygen and limited food. At least superficially it's easy to imagine why sophisticated cognitive abilities are not particularly advantageous in this sluggish environment.

A much more informative gauge (as with our technology) is the actual energy use of brains in different species. That is, after all, an immediate day-to-day burden on an organism, and therefore represents a cost that had better be worth it. One way to measure this is by looking at the metabolic activity of a nervous system (including the brain, spinal cord, and so on) compared with that of the entire body.

That metabolic activity follows a pretty tight correlation. Higher overall body metabolism tends to go hand in hand with higher nervous system metabolism. But there are notable variations. While an elephant's nervous system and body metabolisms land right on the average for their size scale, a whale with a similar nervous system metabolism has a body metabolism almost 100 times higher than the average. Humans actually sit off the curve in the other direction, with a nervous system metabolism almost a factor of 10 above the average for our resting-state body metabolism. That's more than dolphins and a little above that of chimpanzees.

In fact, while most vertebrate species devote between about 2 percent and 8 percent of their total basal metabolic production to their brain, primates have a significantly higher figure. In humans it clocks in at between 20 percent and 25 percent. That means that our minds run with an average 12-watt power consumption for an adult, compared with a total of around 65 watts for our resting bodies. About half of those twelve watts goes toward driving the molecular "pumps" that drive sodium and potassium ions across neural cell membranes, enabling the electrical signals that neurons fire at each other. The other half is from oxygen respiration by those same cells.

Estimating these energy budgets is complicated—and depends on approximations and proxies, like oxygenated blood flow or glucose flux into the brain. Nonetheless, this is clearly an impressive energy consumption for an organ comprising just one-fortieth of our total body mass. It suggests that our brains are definitely worth something to us.

Other research indicates that the human brain operates at, or very close to, maximum capacity in terms of energy efficiency and computational ability, though we should be cautious about direct comparisons between computation in a digital sense and what really happens in a biological brain. There are also hints that brains may balance at the edge of self-organized criticality, on the brink between order and disorder in their coordinated electrochemical properties. It's analogous to the conical organization of a sand pile at the bottom of an hourglass that might disrupt to an avalanche with the addition of just one more grain. Our brains exist in a state of exquisite connectivity but also vulnerable sensitivity.

This maxed-out level of brain operation is directly counter to the popular, and entirely erroneous, idea that we only use a fraction of our brain's capacity. It's not entirely clear where that particular fantasy originated, although neuroscientists have pointed to one likely suspect: a book called *The Energies of Men*, written in 1907 by the otherwise highly respectable American psychologist and philosopher William James (brother of the novelist Henry James and wit-filled diarist Alice James). There he makes a wholly unsubstantiated argument that we utilize only a small part of our potential abilities.

In reality, the effective computational efficiency of biological neurons (as I've described previously) and the human brain is actually extremely high. But how this is achieved is still somewhat mysterious. If we make a more detailed comparison between the properties of digital processors and of neurons, the mystery only grows. A typical cortical neuron is a spindly-looking cell, with its hundreds or thousands of delicate dendritic offshoots and lengthy arm-like axon. It is also quite limited in how fast it can operate. At most a neuron can "fire" its electrochemical pulses about a thousand times a second. While that's certainly not too shabby, a standard modern microprocessor will clock in at *10 billion to 100 billion* "firings" a second.

Similarly, the effective precision of neural firing is not great. A wet, warm brain is full of variability and noise. It's thought that the frequency of a neuron's firing in response to stimuli is a fair gauge of the total intensity of those stimuli combined with the noise, which suggests that a typical neuron suffers from imprecision at the 1 percent level. Roughly one time out of a hundred it will fire inappropriately. By comparison, even a second-tier 32-bit digital processor has a numerical precision of about 1 in 4 billion.

Given these limitations, you'd be forgiven for wondering how on earth the brain can do anything useful. It clearly does—think about the sheer number of factors and calculations involved in catching or throwing a speedy ball, riding a bicycle, or driving a car. Any failure to be accurate, precise, or quick will not turn out well.

A part of the answer seems to be that biological brains work as a massively parallel computational system. They don't wait for each step of analysis to take place in sequence. The thousand or so interconnects between a single neuron and others is a radically different architecture than a standard transistor in a chip, which has three connections to the rest of the world. Neurons can metaphorically (and possibly literally) play the violin while juggling a set of balls. Neurons also employ a complex cohort of neurotransmitter molecules and their molecular receptors, and, as we'll see more of later, modify their interconnects to each other almost on the fly.

Neurons also seem to pool, or take the statistical averages of, hundreds or thousands of noisy signals to overcome the more limited precision of those individual upstream cells. Combining this kind of quasi-digitized rendition of stimuli with purely analog activity (some neurons, like those in our retinas, output continuously graded, variable-size electrical signals) allows brains to make a much more nuanced and mathematically egalitarian accumulation of information and what to do with it. As a result, processing in the brain actually happens quickly and, in the end, with far more precision than the raw properties of neurons would at first suggest.

Critically for the dataome (and we'll soon see why), there once again doesn't seem to be much room for doing better. A complex brain is highly optimized, maxed out, and energetically burdensome. Brain development is super expensive, from the embryonic assembly of neurons (which in humans peaks at a rate of 15 million nerve cells generated per hour during gestation) to the extensive learning period of an infant. And we now even have evidence for fundamental limitations to one of our most defining cognitive abilities: language.

A study presented in 2019, using the tools of Claude Shannon's information theory, concluded that across 14 representative human languages there is a fundamental limit to the data-rate of speech. No matter who we are or what we're talking about, or how fast our language is spoken, on average we transmit information no faster than 39 bits per second. Given the very different encoding strategies that different languages employ, the researchers point to the limitations of human biology as the bottleneck, specifically how quickly we can actually gather our thoughts in order to transmit them.

There is no reason for any species to be walking, swimming, flying, or crawling around with anything more than exactly as much brain as works for them in the broad evolutionary sense, and a brain that is only as efficient as necessary. That our brains seem to be at capacity is not so surprising.

Wouldn't it be wonderful, therefore, if an organism could enhance its cognitive abilities—in memory, in knowledge—without messing around with its maxed-out brain? That does seem to be precisely what our dataome allows us to do. To offload part of our minds, and to vastly increase our memory and information retrieval options, and to do this in a way that persists long after our individual shelf lives.

## Read, Think

It's not just offloading mental capacity that may be important. The very existence of external data and information to interact with creates a unique kind of feedback-induced cognitive development. The dataome can change our brains.

The fact that you are reading this book means that your brain is fundamentally different than if you had never gained the skill of literacy. A research study published in the journal *Science* in 2017 examined the structural and activity changes in the brains of adult humans who learn to read for the first time around age thirty. In this case a cohort of twenty-one women from a region in northern India, who had previously been illiterate in Hindi, were taught a basic level of proficiency over a period of six months, while also agreeing to undergo functional magnetic resonance imaging of their brains. These scans provide structural information on brains, down to a few millimeters in scale, as well as statistics about the level of activity in different brain regions depending on what the subject is doing—reading or not reading, for example.

Remarkably, the researchers not only saw changes in the outer layer—the cortex—of the new readers' brains, but reorganization of deeper structures in places like the thalamus and brainstem. These are regions of our brains that are considered to be old in evolutionary terms, a part of a long-shared mammalian history rather than anything unique to humans. Human literacy is a comparatively new phenomenon, so what is our brain doing, when we read, to parts that evolved much, much earlier?

Some of these changes seem to be connected to better signal synchronization with the brain's visual cortex—perhaps due to the intense demands of visual processing required by reading text. And some of the answer may lie with the spoken word, or at least vocalizations associated with language. Studies show that literate humans are actually better at discriminating certain sounds—learning the phonological rules of language—than illiterate humans. The systems in our brains that evolved or co-evolved with vocalization are becoming modified, even boosted, when we learn to read.

That's not to say there isn't some kind of cost. An idea that the cognitive neuroscientist Stanislas Dehaene has termed "neuronal recycling" argues that a skill like literacy repurposes and develops neurons and the neural circuits they participate in. For instance, Dehaene and colleagues have studied the property of primate mirror invariance. In subjects like macaque monkeys or humans, when we see the left or right profile of an object, or its mirror image, we nonetheless manage to know that we're seeing a single thing—we generalize the object. Our conclusion does not vary when we see the image mirrored or flipped. But in written language this is problematic. In the Roman alphabet for example, a "b" and a "d" can look like mirror images of each other, but it's essential for our literate minds to immediately know that these really are two entirely different things. This means that when we gain literacy skills we somehow undo or override a part of our innate ability at mirror invariance.

Brain activity measurements point to fundamental changes in precisely the parts of the human brain where mirror invariance is handled when literacy is gained. That's actually pretty shocking, because it means we're rewiring (for want of a better analogy) a really important part of our visual cortex in aid of this weird thing called reading. But it actually doesn't seem to impair us; if anything it enhances us. Behavioral tests comparing literate and illiterate adult humans indicate that literate adults are less confused by nonsense shapes, or mirror symmetric letter collections (like "boo" and "ood") because they don't see them as identical. Being literate actually enhances our ability to be flexible in interpreting what we see in the world, training us to seamlessly take context into account.

All of these are fascinating discoveries about how our brains work. But I think the truly astonishing thing is that without a dataome none of this is going to happen in the same way. You have to make symbolic, external, visual representations of language or information in order to have something to train your brain on to see precisely these effects. Clearly that develops for a species over time. This means the structure and function of any modern, literate human brain has to be statistically different from the brains of our less-literate ancestors. But this is not due to any evolutionary change in the human genome. It is a difference due to post-gestational trait changes that are not encoded in our genes, but rather encoded in the dataome. Our external information is literally changing our brains, making them better equipped for a complex world, and for supporting, in turn, the dataome.

## Forgetting

With so much focus in neuroscience on how learning takes place in a brain, it can be easy to, well, forget how brains also forget. But forget they must. At any given moment a species like us is deluged by millions of stimuli. These stimuli induce neurons to build up electrochemical potentials and generate and release neurotransmitters. They also cause long-term physical changes to those neurons. Unchecked, this would be problematic. There is good evidence that much of what our brains spend their time doing is filtering and condensing, to prioritize and sift through this sensory bombardment in order to let go of what is deemed unnecessary.

Recent work in neuroscience has led to the identification of a phenomenon called intrinsic forgetting. A specific subset of cells in the brain seem to be responsible, at least in part, for degrading the physical changes that take place in neurons as memories are made. This degradation, or forgetting, works more like a chronic condition than anything that targets particular memories. To ensure a memory really persists, the brain has to have mechanisms that, in effect, selectively reinforce or preserve the important stuff against a constant erosion.

Our understanding of all of this is still quite rudimentary. There is also evidence that the growth of new neurons can disrupt memory formation, and there is even an indication (from highly invasive experiments on unfortunate sea slugs) that when a memory of a specific stimulus has long since faded, a residual effect remains: a kind of memory of a memory encoded in the expression of the neural cells' genes. It is also likely that different species with different nervous systems, especially across the great gulf between vertebrates and invertebrates, deploy different mechanisms for memory formation, organization, and erasure.

Taken at face value, the forgetfulness of biological brains points toward a perfectly reasonable and necessary Darwinian selectivity. Brains are energetically costly to run and demand specific, costly developmental sequences for an organism. Trimming information, eroding memories, all helps keep the balance between what you need to survive and propagate and the biological cost of more and more neurons and their connections.

But what if you could offload memories and potentially useful information and experiences outside your biological form—wouldn't that be a huge benefit? Your brain can keep on optimizing itself by eroding memories that don't matter so much, but at the same time you can keep a history of useful tidbits and discoveries. Even better, a multitude of signposts and shortcuts can evolve to ensure that external information hits the spot when you need to refer to it. Indeed, from creating indexes to internet searches, we've learned how to bring extraordinary order and efficiency to the wealth of data residing outside of our brains.

## Shortcuts

The graphical representation of data is one quite simple example of how the dataome unburdens our brains and makes them more efficient. This is a form of data management and access that really doesn't seem to happen in our minds. Even our hippocampus, which has been shown to encode temporal and spatial information about the world, does not—as far as we know—contain a literal map. Rather, this little seahorse-shaped package of neurons develops a complex hierarchy of associations, an abstraction of events. And it seems to help tie past and present events together, allowing us to make predictions about the future. But we don't consciously interrogate our hippocampus or experience how it does all of this. By contrast, external information can be sculpted into forms that plug right into our channels of sensory understanding, from visual processing to spatial awareness.

Historically, many of the oldest examples of graphical representation relate to celestial phenomena. The great ring of Stonehenge in Wiltshire in England, which dates back to around 2400 BC, with precursor structures possibly going back to 8000 BC, almost certainly had to do with calendaring the year based on solar positions. In its lumpy gray fashion, I'd argue that Stonehenge is very much a graphical representation, a remapping of projected cosmic phenomena.

But compared with what was happening in the rest of the world Stonehenge was already looking pretty primitive by the time it was built. In India around 3000 BC, members of the Indus Valley Civilization were developing sophisticated written rules to make calendars from celestial observations. In China, good astronomical data was being recorded and charted by at least 6000 BC. By the second and first centuries BC Greek thinkers had built devices like the Antikythera mechanism—some 29 to 39 meshed bronze gears that seem to have enabled calculations of celestial motions and eclipses, and possibly even tracked the four-year cycle of

Olympic games (although it seems unlikely that anyone would have forgotten those).

In the tenth century AD multiple cultures were producing graphical representations and "calculators" for predicting phenomena like solar and lunar eclipses, and even the periodic return of certain comets. A really lovely example is a plot from a tenth-century appendix in a document discussing work by the earlier Roman philosopher Cicero. This graph shows a very modern-looking grid of lines, overlaid with curves representing the motions of the Sun, five planets, and the Moon as time passes. Even by today's standards this is a pretty accurate representation of the data, and a representation that makes sense on a sheet of paper, but is not easy to store in the human brain. Over time we got better and better at doing this. By the nineteenth century we were using detailed statistical graphics, from bar charts to scatter plots. Many of these graphics were works of art in their own right, built to appeal to our aesthetic senses as much as our analytic ones.

In France in 1833 a lawyer named André-Michel Guerry pulled together a trove of data that included crime reports from a newly centralized

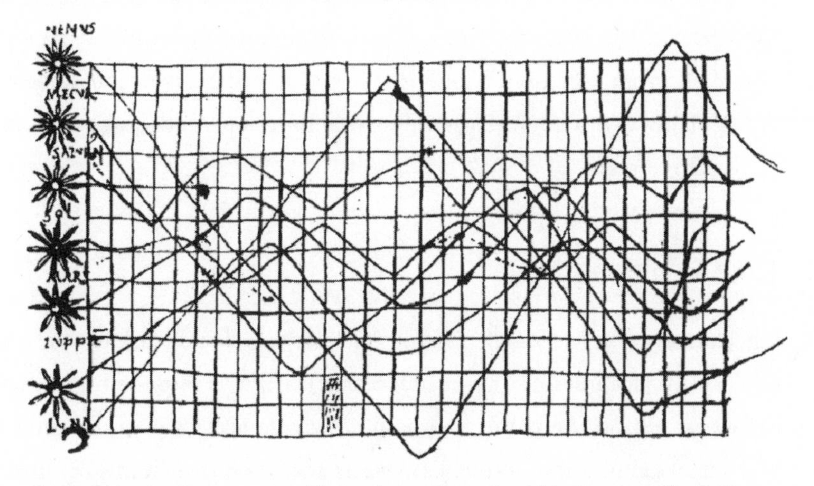

Graph of planetary movements with time increasing to the right, given as an illustration to a short appendix, titled *De cursu per zodiacum*, to a copy of Macrobius' Commentary on Cicero's *Somnium Scipionis*.

national system in the French courts, along with data on literacy, donations to the poor, and even suicides. Out of this he and his collaborators constructed beautiful-looking maps of France, divided and shaded according to quantities like the number of crimes against people or property, or the level of education. Armed with these and other graphics, Guerry attempted to evaluate the "moral statistics" of France. In doing so he helped create the beginnings of modern, empirical social science. Today there is an entire industry devoted to the acquisition and representation of social data, and massive amounts of data to boot, from charting lifespans to dental hygiene. Graphics and infographics infuse our work and our everyday lives, whether through instruction manuals or subway maps and school textbooks.

All of this takes on a deeper meaning if we look at it through the lens of the dataome. The ability to parse complex astronomical data in the form of a graph allows a human to make fast and accurate predictions, a useful thing if you're navigating on a rocking boat or deciding when to plant crops. Equally, to be able to see the overarching patterns and trends in social behavior endows a human with a power of inference that we probably don't have an innate capacity for. Representations of complex information enhance the human capacity for making inspired guesses and leaps of imagination. These manifestations of parts of the dataome arguably boost our cognitive abilities well beyond the constraints of our biological brains.

## Time Machines

Cheeky physicists are wont to say that time travel is quite possible and you're doing it right now. The catch is that it's always into the future, one second at a time. Accessing the past is a whole other story, and even if mathematical physics can invoke esoteric phenomena such as wormholes that bridge time and space, there is no known practical mechanism that could actually take versions of ourselves or our devices back in time. But from fiction to philosophy we're nonetheless fascinated by the possibility. Not only does it make for some movie-worthy adventures and intriguingly paradoxical situations, it would also enable us to have precise knowledge of events throughout human history and the history of the world, from who assassinated JFK to what really happened to the Neanderthals.

I'm being a little facetious, of course. But we know that history is an extraordinarily potent source of knowledge and instruction for the here and now. And we know that because of our dataome. No other species, as far as we are aware, has the ability to store detailed information about the past for its future selves beyond anything but the broadest kinds of environmental and behavioral modifications—whether manifested in old rabbit warrens or patterns of migration.

But the dataome doesn't just unlock the accumulated knowledge of the past. Through our stories and documents, our books, files, and online data, we can share and interact with billions of other humans whom we may never meet physically, or even be aware of in any other way, including humans long since dead. In that sense the dataome is both a time machine and a way to shrink the fabric of space itself.

Furthermore, what we share, deliberately or inadvertently, is not just data or actionable scientific insight into the workings of nature, but also literature, art, and ideas. Among those are examples and prompts for social structure and cohesion, things that may be subtle and unobtrusive but which—somewhere else in the world—can trigger a cascade of new thoughts and actions.

Our tendencies toward prejudice and aggression can latch onto such qualities in dispersible information. That may take an extreme form like the impact of Hitler's awful diatribe on antisemitism and nationalism in *Mein Kampf*. We also see it happen over longer timescales, like our unfortunate and persistently self-serving interpretations of religious or spiritual texts across the ages.

But many, if not most, such linkages are positive, and can be quite subtle. Readings of the great classics of human literature are unlikely to cause violent strife, but can, in gentle ways, bring greater cohesion to humans through their shared experiences. Realizing that someone centuries ago encountered the same questions and challenges that you face is both a potent salve and potential guide. At their best, modern media, from movies to YouTube, can do the same. Recent studies of the behavior of chimpanzees and bonobos even suggest that they form emotional bonds through the shared experience of watching videos together. Think about that the next time you and your companions are slumped on the couch gazing at a screen.

I'd go as far to say that, seen from this perspective, the dataome allows us to circumvent the usual barriers of space and time that a species faces. It has done that from its earliest moments to today, when we've constructed a set of world-spanning systems for communicating and sharing data. As a result, the "mental endowments," to use Darwin's phrase, of humans are indeed continually progressing. But that progression is possible without—it seems—any dramatic biological evolution.

## The Unexpected Past

Some of the most interesting examples of the benefits of a persistent, ancient dataome are cases where data that has lain dormant for centuries or millennia is later discovered to have meaning it never did before. Earlier I discussed cuneiform from Mesopotamia, whose eventual rediscovery and translation helped people develop a better understanding of the nature of ancient myths and their relationship to biblical history. This same region continues to be a focus of modern investigations, with ever more being revealed from its preserved dataome.

Consider the value of ancient human trade records. In 2017 Gojko Barjamovic—a senior lecturer on Assyriology at Harvard University—and his colleagues published a study on commercial records from the Bronze Age in the Middle East. They focused on clay tablets imprinted with cuneiform script originating around 2000 BC to 1650 BC, produced by Assyrian merchants and their trade partners and families. These Old Assyrians participated in a remarkable network of commerce that stretched right across what is now Northern Iraq, Northern Syria, and Turkey.

The data that Barjamovic and his fellow researchers used consisted of 12,000 fully deciphered clay tablets excavated from an archaeological site at Kültepe, ancient Kaneš, located in central Turkey. These tablets are just a portion of nearly 24,000 that still await full modern scrutiny. Most have been found in the archaeological remains of merchant dwellings, and their contents wouldn't seem out of place in the twenty-first century. There are accounting records, shipping documents, business letters, and contracts. In Barjamovic's study some of the text is translated:

> "(I paid) 6.5 shekels (of tin) from the Town of the Kanishites to Timelkiya. I paid 2 shekels of silver and 2 shekels of tin for the hire of a donkey from Timelkiya to Hurama. From Hurama to Kaneš I paid 4.5 shekels of silver and 4.5 shekels of tin for the hire of a donkey and a packer."

That's the kind of business thoroughness you'd not want to mess with. And there are notes that wouldn't be out of place on the desk of a modern CEO:

> "Concerning the purchase of Akkadian textiles which you have written about, since you left the Akkadians have not entered the City; their land is in revolt, but should they arrive before winter, and if it is possible to make purchases profitable for you, we shall buy some for you."

It's easy to find these writings remarkable for their meticulous detail, but that's because we all tend to think we're the first sophisticated generation on the planet. Presumably that's been true for most of human history; put enough centuries between us and we're amazed at how our ancestors could be that smart.

The key point is that these records—these pieces of data that have been in stasis for millennia—have acquired a whole new informational value today. These tablets tell us very clearly that four thousand years ago well-kept data played a vital role in enabling and guiding human activity and behavior. But the tablets also provide us with a probe of exactly what that activity and behavior was, where it was, and what that might mean for us today as our planetary environment changes.

Barjamovic and his colleagues used machine learning word recognition to parse the digitized, translated cuneiform tablets, looking for city names, cargo shipments, and merchant itineraries. Of 26 unique city names only 15 were recognizable, leaving the other 11 as potentially "lost" cities. They then used statistical analyses of the trade patterns to try to establish the actual geographical location of these lost places—and to see whether the sizes and successes of ancient cities match the patterns seen in modern Turkish population concentrations.

Strikingly, many of the mathematical predictions for the coordinates of lost cities seem to agree with proposals from historians' reconstructions of this period. Not only are these locations prime opportunities for archaeological exploration, they can also provide clues to geographical, climatological, and sociological structural changes that have taken place during the human era. Those clues might help us better understand the changes that are inevitably coming down the pipeline for us and our civilization today.

I think this case offers several extremely useful lessons about how parts of the dataome function, and how the dataome reaches back to interact with us. The Assyrian merchants, four thousand years ago, generated and stored their data because it made good business sense. Tracking clients and the flow of goods and income let them run more-profitable operations. But the physical medium of data storage happened to also be extremely persistent. If they had had access to modern paper they might have used that, but would it have survived for millennia afterward? It seems less likely.

The original utility of this information must have been limited beyond a couple of generations in a merchant's family. Yet some forty centuries later the data is resurrected, reintroduced to human culture by being translated and digitized, and actually deployed not only to gain insight to our history but to provide clues to more-fundamental characteristics of human existence. The properties of cities as dynamic, emergent entities are critically important for us to understand as we seek ways to improve the sustainability of our present growth and our planetary environment. That the exquisite bookkeeping of Assyrian merchants in the Bronze Age turns out to be useful for that is a brilliant representation of a phenomenon that is happening all the time, at multiple levels.

There are lots of other examples of modern data resuscitation. The combination of electronic computation, the internet, and the digitization of material has hugely accelerated the possibilities. We now have projects like "Old Weather"—started by the Zooniverse internet platform—which sifts through ships' logs of the nineteenth and twentieth centuries, using online volunteers to tease out millions of data points on ocean temperatures, sea ice coverage, and weather conditions around the globe. Those observations and measurements, even with variable accuracy or precision on the part of the original ships' captains, provide a unique calibration of our planet's changing climate state. Other, similar efforts are digging into the trove of earlier, seventeenth-century logbooks, digitizing hundreds of thousands of scratchily inked pages that can then be pored over by online volunteers.

There are also projects that take the idea of a time machine even more literally. The Digital Humanities Laboratory at the Swiss Federal Institute of Technology in Lausanne has a plan for taking 1,000 years of data on the city of Venice, including maps, monographs, manuscripts, and sheet music, and digitizing and interpreting it. With machine learning, even handwritten material can be automatically translated and the web of interconnections between events and people mapped out in detail. The hope is that there are knowledge treasures to discover, from the behavior of financial markets run under different economic assumptions, to the nature of disease transmission in human populations. Seen from a wider perspective, a substantial part of the dataome of old Venice will be reintroduced, replicated into a new substrate of silicon and electrons.

Perhaps one of the most impactful examples of the re-expression of data—although not previously articulated in this way—is how Charles Darwin became an avid reader of his own writings. His scientific journey was such a slow burn, and his notes and works so lengthy and, well, tedious, that he would often go back to his earlier writing to remind himself of his own once-fleeting ideas.

In doing so he found new meanings and new insights because of the evolving (no pun intended) environment of his thinking. His Geological Diary, kept during the five-year journey of the *Beagle,* is peppered with thoughts and observations layered together, sometimes literally. This wasn't a neat and tidy diary in the conventional sense, but a stack of notes and records on whatever kind of paper was at hand at the time. Here, there, and most everywhere are added ideas, scribblings in the margins, the evidence of revisits and revision, pencil and ink intermingling through the parchment.

It's hard to see how this kind of iterative self-referencing can happen without permanent data storage external to our minds. It's true that you might tell another person your ideas and then talk to them again later and have them remind you of your thoughts. That's undoubtedly a valuable mechanism for developing insights. Or, as we've seen, there are techniques of oral tradition that are extraordinarily robust. But these cases involve specific, tuned, curated sets of information. Data that is generated in the moment, as notes or spontaneous ideas, is likely to be held imperfectly if it exists only in human minds. It will be prone to change in many ways, including suffering meme-like or virus-like transmission and mutation.

Self-referencing can be shocking. Haven't you, at some point or another, come across a letter or email you wrote years before and been struck by how it feels almost like reading another person's thoughts? We all change with time, as we learn skills and gain knowledge. The old lament of "if I'd known then what I know now" is another example. Yet the dataome *does* allow us to learn from our past self, and from other people, with a specificity and accuracy that seems hard to attain in any other way.

There is a striking resemblance, as I'll show below, between how data from the past takes on new forms and new utility and what happens in biology itself. I'd be very hesitant to claim anything more than the appearance of similarity, but I also think there could be something tremendously important in this pattern. When we study the evolutionary history of species and their underlying genetics, there are many examples (and these may be the norm more than the exception) of how genes acquired in the distant past can play unexpected roles in the future.

## Lineage

The young of the glorious monarch butterfly love gobbling up plants called milkweeds. In the larval, caterpillar state, these are the monarch's only food. But milkweeds are full of toxins such as cardiac glycosides, evolved via selection specifically because—we presume—they help ward off hungry insects.

Recent research suggests that the monarch has three genetic mutations that have enabled it to evolve over time to consume milkweeds, and even to fill its body with the plant toxins in order to, if we're blunt, make predatory birds puke. These mutations appear to have occurred one after another, but it's only the full, combined effect of all three that confers the full benefit on a caterpillar. If they only had the third mutation and not the previous two, they might actually have other dysfunctions. But it's not as if those earlier mutations could have known their future role for a third, as yet to come, mutation. Rather, it seems that those past genetic mutations take on a new meaning, serving an additional purpose, not unlike elements of the dataome.

On a more abstract, general level, there are the phenomena of lineage selection and so-called "neutral" mutations in evolution. Lineage selection refers to the idea that certain biological traits may not make a species particularly fit at that moment (in the evolutionary sense), but these traits offer the most options for *future* evolution. The scientist Ilan Eshel, and later Richard Dawkins, are among those who have previously brought this idea up. A hypothetical example is that the first animal with a segmented body plan—repetitive segments, like in modern millipedes—might not have been particularly fitter as a result, but its descendants could've benefited enormously from the flexibility of options this enabled. The lineage survives because it turns out to be more *evolvable.* Therefore, the evolution of evolvability is important.

Some evolvability is very stealthy, though. Neutral mutations are changes in genetic code (perhaps simply a single changed nucleotide letter) that don't alter anything about the organism—or at least no trait that affects the survival of their lineage at that time. But an innocent neutral mutation could, in principle, lay the groundwork for future mutations that, in combination with the earlier neutral change, produce significant change and adaptability in an organism. Those benefits might be felt millions of years later, and in retrospect the neutral gene mutation was actually a super high fitness change. It's a little bit like if you'd always had a really good sense of balance, but that didn't give you any big advantage in life—until the day that electric unicycles became the world's only form of transportation.

A dataome may open a window to even more evolvability for a species. In biological systems, if a trait is weeded out by selection because it doesn't confer enough increased fitness, it can simply go away forever.

The gene variants that are critical to the trait will disappear from the population in our standard picture of things. For the dataome that can all play out differently. The example I used of ships' logs is a good one. If the keeping of laboriously handwritten logs on large sailing ships was the same as a biological trait, it would have disappeared from existence by now. But the dataome, in this instance, preserved both the outcomes of that trait and the knowledge of their production. We could, in principle, easily retrain ourselves today to replicate the way that these old ship logs were kept. There doesn't seem to be a way for that to work with genes. I can't reacquire an ancient gene once it's lost. But I could, with data on paleogenomics and gene-editing tools, perhaps recreate it—in other words, with help from the dataome.

The dataome can also store contingency plans and instructions to be applied in situations that may or may not ever challenge us. We've all at some point acquired an object that comes with an instruction manual. Beyond giving us information on how to operate or use our new acquisition, whether a TV or a spaceship, the manual inevitably has the comfortingly labeled "Troubleshooting" section. Some or all of the troubleshooting scenarios may never, ever occur in the history of our species, but here they are, just in case. On a much grander scale we have evacuation plans for cities, or emergency procedures for fires, earthquakes, or floods. We have recorded constitutions for countries, and medical libraries for all that can afflict us now and in the future.

Many of these plans and instructions are themselves the product of earlier evolutionary lessons. No one automatically knew the optimal rules for escaping a burning building until people had experienced that trauma. But we're pretty good at imagining and modeling futures that we may never encounter, and being prepared for them. In computer science, for instance, we sometimes generate what is called a lookup table. A lookup table might be information such as the results of a computation that is exceedingly intensive to perform, but rests on the starting value of perhaps a single variable. Instead of computing the outcome on the fly, which would slow down a piece of software, we can compute all possible outcomes beforehand and simply store the results, indexed for rapid retrieval. In a very real sense I think this is a form of engineered evolvability.

Because biological evolution prunes out traits and their genetic roots, biological lookup tables are much more limited. It is unlikely that a quick "lookup" editing of my existing genes could turn me into an octopus, even though we did have a common ancestor some seven hundred million years ago.

## Data Holes

Another way to try to gauge the benefits of a dataome is to examine instances where it has been damaged, altered, or made unavailable to us. If that happens you'd expect to see some negative consequences. Luckily there are quite a few such examples, because we are often prone to behaving in ways that are, not to sugar-coat it, sublimely idiotic. At the start of this book I discussed how the Sassanians purged Babylon of its scribes and cuneiform libraries. Popular history also talks about the accidental burning of the great library of Alexandria by Julius Caesar in 48 BC when he was besieged and trying to burn boats to clear an exit from the city harbor. While the library survived as an institution, the sense of loss of tens of thousands of scrolls (real or imagined) has echoed down our timeline and has been variously accused of holding back the development of human knowledge for centuries.

Much more recently, the Cultural Revolution of China from 1966 to 1976 didn't just launch an astonishing number of determined efforts to rewrite the history and social order of the country (at enormous human cost); it also went after literature, art, and science. Thousands of cultural heritage sites were destroyed or damaged. Manuscripts, books, and artwork were burned. Living writers were often persecuted if they didn't adhere to mandated guidelines, topics, and perspectives.

When a booklet circulates called "Four Hundred Films to be Criticized," you know that something bad is going down. Very few new major movies were made in China during this period at all. On the science front, most research came to a screeching halt except for some military work. Again, the human cost was huge, with a whole generation of young, scientifically inclined minds effectively quashed. As a result of all of this, together with strict controls on what could and couldn't come in and out of the country, China tore a huge hole in its dataome and felt the consequences in its post-revolution economy and the education of its people.

Throughout human history there have been new orders, new regimes that march in and declare the previous systems of governance and knowledge obsolete or, worse, heretical. And that is perhaps the most damning and certain evidence for the intrinsic value of the dataome: the persistent recurrence of humans destroying each other's data repositories. They're seen as potent ways to promulgate a culture, and if you disagree or want your own power, the first thing you do is try to erase that culture's knowledge. Burn books, pull down statues, outlaw ideas. The fact that we recognize or sense the power of knowledge is unarguable, even if it's not always clear what losing specific repositories of knowledge does to the survival of our species as a whole.

## Progress

Up to this point I've only dropped hints about where these observations and ideas about human data could be leading us. But in thinking about the ways that a dataome benefits us—in the greater, evolutionary sense—we have actually already crossed a line. That line is a conceptual boundary, and is about establishing where "we" end and the dataome begins.

It is certainly still possible to see external data as simply a part of the environment we inhabit—an environment that includes all other humans, the other species of Earth, and the inanimate but decidedly active world of flowing air, sea, land, geophysics, chemistry, and energy. Superficially, what we've done with our external information could be equated with a bird building a nest, or termites building a complex mound. But those structures don't come close to the complex, dynamic, and intimate properties of a dataome. They do not reshape the physical or informational environment to such a degree. Nor do they demand as much from their biological makers.

Modern science has taught us that Darwin's tangled bank is a lot messier than he could have known. As we'll see, phenomena like horizontal gene transfer and microbial coexistence in the microbiome hugely complicate our picture of life. Discoveries of recent decades such as the single-celled archaea, the deep hot biosphere, and the strange and genome-rich giant viruses all raise urgent questions about life's true trajectories across time. And the existence of epigenetics complicates and even challenges fundamental aspects of Darwinian selection and evolution.

Nonetheless, no one phenomenon has really usurped its central story: the diversification and pruning of lineages. Or, phrased differently: the endless experimentation and attrition of qualities in biological systems, like an urgent bubbling soup that constantly threatens to boil over but inevitably self-corrects, because otherwise there would be no soup to observe. Within that framework there are hints of further rules for what types of structures can emerge, and why. Writing in 2011, the evolutionary biologist and theorist David Krakauer beautifully articulated some of our cutting-edge thoughts on the fundamental nature of evolution and life. He makes the case that at a higher level of abstraction, above isolated cells and molecules, it seems that life is driven to complexification. Organisms, especially those with more complicated machinery—such as multicellular life—appear to be selected to further increase the amount of information they encode about their environments, and about the other organisms that make up part of those environments (the meaningful mutual information we saw in the context of Shannon's entropy). Networks of cells become capable of extracting the information around them, and using it to infer sources of energy and predict the behavior of other life. These talents come from the observation (for want of a better word) of regular or predictable patterns of the past—patterns then redeployed for guessing the future.

These ideas seem to be getting at the real implications of Darwin's description of "progress." Not necessarily toward a conventional idea of perfection, but toward more and more-meaningful information, and better and better projection. Consequently, if data and its information are really what determine life's function and trajectory, the line between us and the dataome is most definitely blurred, if not erased outright.

This next piece of the story involves taking a deeper look at precisely that function and trajectory, at the complex and fraught world of evolutionary biology and the nature of life itself.

# 5

# GENES, MEMES, AND DREAMS

*Imagination is the Discovering Faculty, pre-eminently. It is that which penetrates into the unseen worlds around us, the worlds of Science. It is that which feels and discovers what is, the real which we see not, which exists not for our senses.*

—Ada Lovelace, 1841

Life is an unwieldy beast for polite study. It's a phenomenon operating at an astonishing number of levels. When Darwin slowly assembled his grand proposal on the origin of species, he was in many ways still privy to just a narrow range of life's properties. He didn't have a molecular basis for life to refer to, and the laws of biological inheritance being worked out in Gregor Mendel's experiments on pea plants wouldn't be widely known for another several decades. Nor did he have modern insights into the nature of developmental biology—the assembly of embryos and the toolkits of genetic regulation. He also (although he came close) didn't have a complete view of life as a set of global chemical networks and systems nested inside other networks and systems, all grinding away for billions of years.

Today we understand somewhat more of life's inner workings and its trajectories across vast stretches of time. Living things are assemblies of atoms, molecules, macromolecules, cells, populations, interactive entities, predictors, and informational inference engines. Life exists in environments and makes environments, hashing and rehashing this rocky planet as it orbits around a middle-aged star. The presence of life here has dramatically rewritten Earth's surface chemistry, and modified its hydrological processes—from oceans to clouds. Life has influenced the meanders of rivers and the erosion of continents, and it's ridden out cataclysms and the inexorable changes that come with time, be they sporadic asteroid collisions or the steady brightening of a thermonuclear sun.

It's on these edges of scale, microscopic and planetary, where some of the most intriguing patterns appear. Because it's here where it's most obvious that the restless experimentation of living systems locks onto certain rules, certain solutions or recipes for overcoming the challenges of existence. We now know that across all life on Earth there are effectively ten core metabolic tricks employed by species. Some are familiar; we ourselves make use of oxygen respiration, and plants use oxygenic photosynthesis. Some microbes use fermentation, converting complex molecules like carbohydrates into a variety of simpler forms like alcohols or carbon dioxide. Other microbes use nitrogen fixation. There are also less-familiar pathways, like methanogenesis, iron respiration, or sulfate respiration. All of these processes—the details of which can be horrendously complicated—shuffle elements and electrons around the planet in a vast and interconnected network of energy flow.

But although these fundamental metabolic processes have been chugging away for a very, very long time, the specifics of their implementations are diverse. The methods of the recipes may have been passed on through billions of generations, via many genes that are remarkably unchanged, but the chefs and sous-chefs are never quite the same, nor are their kitchen practices. There are sulfate-reducing microbes in your gut right now. But there are also sulfate-reducing microbes thriving off scalding hydrothermal water plumes at many kilometers' depth in parts of Earth's oceans.

This simple example hints at something deep. When we look at life in bulk—as a planet-wide, interrelated phenomenon—what matters most are those recipes or solutions, the *ideas* for life. And just like the ideas held in our minds and in our dataome, while they can be traced to a physical rearrangement of a piece of the cosmos, they also, in effect, exist outside of this. Organisms and even genes are simply convenient temporary parking places.

That is a profoundly unsettling but beautifully useful way of looking at the world. As we'll see, it springs from a multitude of developments in the last century or so in how we conceptualize and track life and its evolution.

## Kin

To take a first step through that door let's carry out a small thought experiment. It's a critical example of the interplay of information and evolution, and it involves the challenging (and quite contentious) subject of the behavioral pull between closely related individuals and the social groups that they do, or don't, belong to.

Imagine, if you will, that you're one of half a dozen siblings who live in a small village, we'll call it Genesville. Options for work and for partnership are pretty limited in Genesville. Consequently, you've ended up single and the default babysitter and support-on-call for your brothers and sisters with children. Pretty often this eats into what few opportunities you have to either start your own family or further your career. But in many respects, you're okay with this. It strengthens your ties to your siblings and, to be honest, it feels pretty good to be able to participate in their child-rearing without experiencing the full burden of keeping a family.

While this might sound like the premise for an earnest but deliriously tedious novel, it actually reflects a critical set of observations about the evolution of altruistic behavior. Your seemingly selfless actions are equivalent to those seen in numerous other species. But why should this be so—what on earth would drive individuals to sacrifice aspects of their own lives and to forgo opportunities to propagate their own precious genes?

Suppose that you can put on a pair of goggles that allow you to see the ebb and flow of genes across time in Genesville. You can see your parents and the fifty-fifty random mix of their genetic material that funnels into you and each of your siblings—half from your mother, half from your father. You can see the flow of genes from your siblings to their children. And you can recognize that your nephews and nieces—on average—share 25 percent of your own genetic material. That all adds up, and as a result, without directly reproducing yourself, you are nonetheless part of the guardianship of the future of a majority of your own genes.

You can also, with the goggles, see that in order to propagate an equivalent amount of your personal genetic material you'd need to have multiple children of your own—an exhausting proposition. So, your current life as part-time caregiver and cheerleader doesn't seem so bad after all. In a world driven by genes your altruism has a completely reasonable explanation. Your behavior benefits the genes that you and your closest relatives are carrying into the future more than if you had one or two children yourself. The genetic benefits of your actions outweigh the costs of not being a parent.

Except, of course, in real life none of us possess those magical gene-viewing goggles. Furthermore, the kind of behavior I've described—siblings helping siblings raise offspring—is seen in species with wildly different cognitive abilities, from squirrels to birds and more. How then do we, or other creatures, know that our behavior actually makes the best sense for propagating our genetic material?

The answer is that we don't. Neither do the squirrels or birds—

although there can be chemical signals that enable recognition of relatedness (families may smell similar). But these behaviors increase the overall odds of genetic continuance, while alternative behaviors don't. Consequently, the predisposition to those alternatives disappears over time. That generational pruning is, again, at the core of natural selection, and an even longer-recognized mechanism. As Darwin pointed out, back in the fifth century BC the Greek thinker Empedocles considered life to consist of random structures, of which we simply witness the ones that survive.

If, in my above hypothetical example, I was instead an unstable, unreliable, self-absorbed individual, I wouldn't participate in the raising of my nieces and nephews. Although they might not suffer much from my absence, in the long term there could be subtle disadvantages that will pursue my lineage through time—all because I spent too many hours at the pub.

This kind of story is captured in a concept known as *kin selection,* and most famously illustrated through a deceptively simple-looking mathematical expression called Hamilton's Rule. The rule, first described by the evolutionary biologist William D. Hamilton in papers published in 1964, draws together theory and data to propose that the altruistic behavior I've described is directly correlated with the degree of genetic relatedness of individuals. Most strikingly, this formulation also seems to provide an explanation for the kind of divisions seen in societies of insects like bees or wasps—with reproductive queens and hordes of sterile workers. In these species the way that genes are passed down is fundamentally different than for creatures like humans, and that is reflected directly in their altruistic and social behavior.

What is quite remarkable is that gene-driven kin selection makes sense if it is solely the information encoded in genes that really matters, not the specific *way* that it's encoded. That information shouldn't care about chemistry or who the vehicles are. It will replicate in any way that works, simply because if it doesn't it ceases to exist in the world. This is a high conceptual bar, and it leads to some of the most contentious and still-unresolved conundrums in evolutionary biology. Those conundrums will in turn lead us to a new way of looking at the dataome and its relationship to our own molecular code.

## Selfishness

In 1976 Richard Dawkins published his popular science book *The Selfish Gene*. This book wasn't just a best seller; in 2017 it was listed as the most influential scientific book of all time by a Royal Society poll. While one might (jealously) question how any book can really be singled out for such an accolade, there is no doubt that *The Selfish Gene* is a great read, and has provoked an outsized amount of debate and reflection over the past decades.

At its core is an argument for an entirely gene-centered view of evolution, drawing extensively on earlier work by Dawkins himself and scientists like William Hamilton (he of kin selection) and the biologist George Williams. From this perspective the primary components of heritable information, encoded in pieces of DNA, flow around the world, subjected to differential survival, and uncaring of where they are contained at any moment. In this picture, the processes of natural selection make the most sense from a viewpoint where genes are "replicators"—entities competing at the root of Darwinian evolution, carried through the world in survival machines, or "vehicles." Those vehicles might be an organism or group of organisms that share in the success or failure of their replicator passengers.

All other parts of the living and non-living world are just an environment in which the information represented by genes propagates—an environment produced in part by the actions of the genes themselves. But genes are not purposefully directing this theater; they are participants only because their properties happen to result in their continuing presence, and the presence of the information they represent. Dawkins has himself said that using the term "selfish" led to confusion. This was never meant to suggest that genes are thinking or exerting some kind of will on the world, but rather that we might see the results *as if* genes were selfish. As an example of a metaphor that is so compelling and easily digested, yet prone to sending our thinking onto a narrow path, this is a pretty good one. We'll come back to that particular point later.

There are, naturally, many caveats to this seemingly elegant, unifying idea. There is still intense debate today about what the fundamental "unit" of Darwinian selection really is for life: whose relative fitness is really being pitted against the world. Is it indeed that of the gene (the replicator), as Dawkins's book asserted? Or is it the phenotype of the gene, or is it the organism (the vehicle) containing the gene, or is it still higher up the chain in the form of groups of organisms?

Other scientists, including, notably, the hugely influential biologist E. O. Wilson, have advocated for the critical importance of groups and societies, so that *group* selection has to be studied. In that case, fitness and selection applies to social groupings—one family versus another—or between entire genomes, not just genes themselves. That possibility can be further extended to the idea of multilevel selection, where Darwinian selection takes place across different scales and degrees of interaction (including through competition and cooperation), and there has to be a simultaneous weighing of attributes, with no single unit of selection, but rather sets of cohesive systems of selection.

Another complication is that genes themselves can play fast and loose with some of the rules we come up with. We know of "supergenes," great big clusters of genes on a single stretch of DNA that gets inverted, literally flipped backward along the molecule, preventing the usual genetic mixing up of recombination in sexual reproduction. As a result, the supergenes seem to ride out generations, evolving as if they were a single gene, and facilitating the rapid development of specialized, complex phenotypes. For some species, like zebra finches, mountain honey bees, and ruffs, the supergenes have freed up evolutionary strategies from some of their usual constraints. For the ruff, a wading bird, the supergene has allowed three "types" of males to develop, together with mating behavior that would shock attendees of even the most decadent masquerade balls.

Other properties of genes can highlight their "selfish" qualities. In the late 1920s and early 1930s it was discovered that bacteria can exchange genetic material, a process we now call horizontal gene transfer. This occurs through a variety of pathways: microbes can uptake genetic snippets that have been released into the environment (a process called transformation), or via the insertion of foreign code by the actions of a virus (transduction), and sometimes through a physical cell-on-cell bridging (conjugation). In single-celled bacteria and archaea horizontal gene transfer is rampant, confounding efforts to construct evolutionary trees, but is clearly a mechanism strongly favored by evolution. Evidence has even been found in primates and humans of small numbers of genes that are "foreign" in coding and likely the result of some kind of horizontal transfer. Yet not all horizontally exchanged genes confer benefits, or have any function; some may be genuinely selfish—hitching a ride while they can.

There is also no easy one-to-one mapping from the full DNA sequence of a species to all of its biochemical and structural characteristics. For instance, there are transcription factors in the form of proteins that control the rate at which DNA is transcribed into messenger RNA inside cells—an essential part of gene regulation, so that the right stuff gets made in the right amounts at the right time and place. In humans there are at least 1,600 of these specialized molecules that impose themselves on whatever genes an individual's DNA encodes. It's like having a forceful orchestral conductor specifying precisely when each player gets a turn. That suggests that we should also ask whether the "conductor" is itself being judged on its performance: Is it too undergoing selection?

Rounding out this list of caveats is the phenomenon of epigenetics: the heritable modification of gene expression that doesn't alter the coding of DNA itself. Some chemicals, like so-called methyl groups, can bond to parts of a cell's DNA and inhibit genes there from being transcribed.

But the location-specific presence of these chemicals can actually be passed between generations, and seems to correlate with factors like the external environment or even age. Trauma and stress responses in individuals can leak across generations through epigenetic changes. Your parents' trials may, unfortunately, be part of your trials too.

The bottom line is that it remains surprisingly challenging to agree on how to characterize units of natural selection, and on how nature actually weighs what we've conveniently labeled as fitness in biology. A fully gene-centered view of evolution is unarguably a powerful and provocative way to examine the world. Like the dataome, it is a highly illuminating lens to peer through. But evolution is also a great example of how science is a perpetually unfinished book.

One fact does seem to hold fast, though, and that is how the propagation of information, by genes and genomes and what they do to the world, is welded to the heart of evolution. The dataome also propagates information. So, how tightly bound together are our genes and our external data?

## Infectious Ideas

In an effort to provide another example of the phenomenon of selfishness, Dawkins gave a name to the now familiar concept of memes, which I briefly mentioned at the start of this book. These "mind viruses"—to use Dawkins's provocative description—are ideas that are not just readily spread but can also induce new behaviors in their carriers. Indeed, the spreading of a meme is itself an induced behavior, whether it's through a human conversation or a share on social media.

Dawkins's term crystallized thinking on a phenomenon that had long piqued people's interest. Back in 1880 Thomas Huxley (known as "Darwin's Bulldog" for his support of evolutionary theory) wrote "The struggle for existence holds as much in the intellectual as in the physical world. A theory is a species of thinking, and its right to exist is coextensive with its power of resisting extinction by its rivals."

Memes can also act as if they are selfish, because sometimes they're detrimental to their carriers. Humans are prone to becoming obsessed with ideas that can lead to disadvantage or even death. Starving artists, impassioned protestors, religious zealots, thrill-seekers, and political ideologues can all seem to be on a course to self-destruction because of ideas that they nurse and propagate throughout the world.

To explain these seemingly irrational patterns, we can say that the memes, or ideas, are simply using their carriers in the way that biological viruses hijack their hosts, or that genes use organism-based vehicles. Human minds are a landscape in which ideas can propagate and compete with each other, following rules that strongly resemble those of natural selection. What happens to the humans, good or bad, is mostly of secondary importance to the continuation of the information, its further replication.

This vision is intriguing, disturbing, and enormously contentious. To this day it's borderline unacceptable in many scientific circles to treat memes as anything worthy of scientific analysis beyond their bearing a superficial similarity to what happens in biology (and to be clear, Dawkins never really suggested otherwise). This is especially true when it comes to ascribing mutualism to memes and genes—speculating that the evolutionary fitness of genes might be impacted by memes, and vice versa. That hasn't stopped a lot of ink being spilled on memes (filling up a corner of the dataome, with some irony), with certain scholars proposing formalisms of so-called memetics, and a central role for them in cultural evolution.

I'm not going to jump very far down that particular rabbit hole here. The main reason for scientists' conservatism toward memes is that it is enormously difficult to separate out cause and effect in a complex, intertwined, messy set of systems like life and minds. Finding the phenomenon at the root of things, the fundamental actor, is supremely difficult. That doesn't mean that a simplifying approach, or a universal rule, can't be the answer. But proving that to be true is why most scientists still have jobs: it's a long road.

With that cautionary note in mind, there is such an appetizing resemblance between the notion of replicating, evolving information encoded in genes; the existence of memes; and the characteristics of the dataome, that we have to take a look.

Previously I've said that I don't think the dataome is just a collection or consequence of memes; instead, memes represent a subset of entities working across the border between the dataome and human minds. A popular catchphrase will bounce back and forth between minds and dataome. By contrast, a bus ticket or a database of winter cloud cover in Belgium, while definitely a part of the dataome, probably doesn't spend much, if any, time in human minds.

The dataome also amplifies memes and aids in their survival. In a human culture, beliefs or values are more easily shared and resilient because they exist as commonly accessible information—in physically manifested data (like the Quran, the Bible, the Vedas, the Tripitaka, the writings of Karl Marx, or Hobbes's *Leviathan*). Memes have more access to hosts and hosting media in a species with a dataome. Therefore, the better that dataome is—in ease of access, efficiency, larger size—the better it is for those memes. There are intriguing similarities between this arrangement and the arrangement of genes and organisms. As we'll discuss shortly, a gene can't go it alone in the world. It both relies on and contributes to the entirety of a biological system, be it a cell or a population of a species. The better those biological systems work, in terms of reproduction, repair, and diversity, to withstand changing environments, the better things are for the genes.

Today, in a way that has not really happened in the past, the information represented by genes also finds itself represented in the dataome. For instance, a very stable set of genes in terrestrial biology are those that code for some of the structures of ribosomes in single-celled organisms.

Ribosomes are large molecular machines that are vital to the production of proteins. Consequently, these genes and sections of their codes haven't evolved much at all over millions, even billions of years. A particularly well-studied set are called 16S rRNA, and thanks to genomic laboratory analyses we have decoded thousands of 16S rRNA gene sequences from different species. Those reams of data now exist within the dataome.

In other words, the information represented by 16S rRNA genes has found its way into an entirely new storage and replication system—that of books, electronic media, and countless computers and data servers across the planet. You might object that this has no significance—the 16S rRNA information is no longer really doing anything, it's not resulting in new ribosomal molecular machines that churn out proteins in the world. It's not exerting its original capabilities. But the point is that, in the framework of selfish genes, those outcomes were *never* anything more than a means to an end. If the sole reason for the existence of genes is that they *can* continue to be, to exist in the world, then whether the information they represent sits in an organism or in your hard drive doesn't matter.

Of course, the dataome might struggle to continue to exist without its biological minders. In that sense, the original function of 16S rRNA in the organic world is still critically important. But now so too is its function as an object of intellectual curiosity for human minds, for scientific research, and perhaps for future genetic engineering. All of which select it for maintenance and replication within the dataome.

There's an argument to be made that none of this should be surprising because the processes of gene replication in biology, and the ways in which genes actually evolve, are already far from simple.

## Replicator Blues

There is debate, if not outright dissension, over the notion of genes as straightforward replicators. As I've mentioned, the physical traits, the phenotypes, due to a gene are the result not just of what's encoded in its nucleotide sequence, but also of the way the products interact with the whole environment of other genes, biochemistry, and the external world.

Most critically, a gene held in a stretch of a DNA molecule is certainly not capable of strict self-replication by snapping its microscopic fingers. It takes a whole village of other molecular machines acting through the larger process of cell division (or mitosis) to replicate that DNA. And even though all those machines are themselves encoded in DNA, they don't just spring unaided from it. The weighty industrial complex of Earth's biology, bootstrapped into existence across billions of years, is essential.

Fundamental, or nearly pure, replicators do exist in the world. A very simple example is a crystal that grows by adding atoms to itself in a way that precisely replicates the "base" arrangement of the atoms already in the crystalline structure. Then there are autocatalytic chemical reactions in which a molecule directly causes more copies of itself to be produced. The formose reaction is the most famous of these, in which products of a chemical reaction can themselves engage in the next cycle of reactions to produce more of those same products. Scientists such as the chemist Julius Rebek and his colleagues have also managed to construct synthetic molecules that can act as the direct template for making more of themselves. Critically, though, all of these examples are considered somewhat trivial, in the sense that they're passive. The information contained in the thing being replicated is only marginally more than in the bits and pieces being assembled. The sum of the parts is, well, pretty much just that.

We can also create software replicators quite easily. I can write a short piece of computer code whose sole function is to *compute* and write out an exact copy of itself (without simply referring to its own code or any other input). In computer science such things even have a special name—they're called "quines." The term was coined by the polymath cognitive scientist Douglas Hofstadter in 1979, but the idea of self-replicating machines or software has a much longer history. Here too, external machinery remains essential, in the form of an operating system that interprets the quine and enables it to push electrons into the right places to remake itself.

That leaves us with a conundrum. Genes clearly do propagate copies of themselves into the future—they replicate. But not only does this involve a bunch of other molecules and a complex hierarchy of systems, it involves host organisms (to use that term loosely), species, and ultimately the entire network of a planetary biosphere. At the same time, we can explain critical elements of Darwinian selection, such as kin preferences, by invoking the genes as if they were direct replicators.

To add to the confusion, contrary to a long-held assumption that new genes only emerge when old ones are blended together, or are accidentally duplicated or broken up, there is now evidence that some new genes are built quickly from scratch.

Starting in the mid-2000s, a number of researchers began to speculate that some of the genes they were seeing in species such as fruit flies were extremely young in evolutionary terms. These genes looked kind of unpolished, as if natural selection had not yet smoothed out the rough edges of their function. Furthermore, these "*de novo*" genes seemed to start from sections of DNA in between the known codes of genes—those vast noncoding stretches that don't usually produce anything, sometimes called "junk" DNA (although that term is slowly retiring, since little is really junk in these polymer molecules, and they may be remnants of viral genes and earlier evolutionary experiments).

We now know that these young genes pop up all over the place. Another good example is the antifreeze gene in Atlantic cod, a DNA sequence that produces proteins that inhibit the destructive growth of ice crystals in the fish's cells. There evidently isn't a deep lineage for this gene; it seems to have appeared *de novo.* In one Asian rice subspecies, there look to be as many as 175 genes that were created *de novo* over just the past three to four million years.

How this happens is still not entirely clear. But somehow, stretches of noncoding DNA acquire the critical start and stop sequences that communicate to a cell's machinery to transcribe this piece of DNA. These stretches of DNA must go through a lot of quick selective shakedown, too, as previously noncoding DNA will be full of nonfunctional sequences, those rough edges. But many of these previously quiet stretches of DNA show evidence for "useful" codings: groupings of nucleotides that correspond to potential protein molecules. The yeast genome is crammed with hundreds of thousands of sequences that look like they should do something useful, but for now, don't. It's estimated that *de novo* genes could represent a tenth of the active genes we see in species. In other words, it's not just replication that matters for genes, it's the dynamic flexibility of the entire genomic system that enables gene creativity—from well-known methods of duplication to upstarts like *de novo* generation.

These insights reinforce the essential observation that genes (selfish or otherwise) are not completely fixed codes. They can and do change over time, and their expression into proteins and traits is seldom a simple, linear process of mapping A to B. They all look more like temporary manifestations of experimental solutions to self-propagation. Genes, as written in molecules, are an instantiation of a phenomenon, something agnostic to its representation in atoms or symbols, but nonetheless propagated in the world.

That more nuanced way of describing things is very similar to what we've seen of how the dataome operates, with information held, replicated, resuscitated, and propagated in a multitude of ways. But could the dataome effectively have its own units of selection, or mesh somehow into the still unresolved question of what biological selection really acts upon? To get to the bottom of that we have to backtrack even further, and consider how we conceptualize what's wriggling around in front of us in the first place.

## Organism (Part I)

What's in a name? Quite a lot, in fact. Science regularly utilizes categories, labels, concatenations, and shortcuts to try to smooth its way to doing business. Newly recognized phenomena need unique specifications to be referenced accurately, and terminology helps connect discoveries to what has been studied previously. Meanwhile human languages constantly evolve, sometimes accommodating that terminology, sometimes forcing it to keep up.

But there's a massive catch. As much as we might like to behave differently, when we latch onto a word or phrase to describe something in nature we inevitably buy into any limitations or inappropriateness of that description. We pick up the associated meanings, whether intentional or not.

This goes beyond basic naming conventions. Science constantly uses metaphor and analogy. These are critically important tools for helping us conceptualize phenomena that may be far removed from our individual sensory experiences. I've done this with abandon already throughout the pages of this book. But labels and metaphors are like a razor-sharp double-edged sword (metaphorically speaking). They can also predispose us to thinking a certain way; and in some cases that can severely obscure our view. Atoms are not really like bouncing billiard balls, and DNA is not really a blueprint for life—which would erroneously suggest a simple one-to-one correspondence between the composition of that organic polymer and all the attributes of living things.

Consider the example of the term "organism." Originally the word *organon* was used by Aristotle to refer to a part of a living being (particularly a human). But later in Europe, during the 1600s, this usage evolved to the word "organism" from the Greek ὀργανισμός, or "*organismos,*" and was being applied to specify an organic structure or organization. For example, you'll find references to things like the Organism of the Eye—meaning the structure of the eye. Then, into the 1800s the usage of organism morphed closer to its modern meaning, indicating a whole living animal or plant, a body exhibiting organic life and structure. "Organism" became a catchall term for anything that amounts to a living unit, specifically what appears to be a self-organizing and self-actuating entity.

But think about that for a moment. You would likely take a look in the mirror and call yourself an organism, made up of pieces like a brain, heart, lungs, liver, bones, and skin. Yet those human pieces are themselves multicellular things, composed of perhaps 30 trillion or more cells in total, together with at least as many single-celled microbial passengers. Each of these cells can help regulate itself and, in principle, reproduce itself (although not all human cells do). But we don't usually talk about our human cells as each being an organism, even though these cells themselves contain complex, independent structures, and we continue to talk about a single-celled bacterium as an organism. As the Nobel-winning geneticist Barbara McClintock once said, "Every component of the organism is as much of an organism as every other part."

Admittedly, one of my cells might not last long if removed from its usual place, but in principle it could be sustained in a laboratory. If we can't call this cell an organism then what exactly should we call it? Our history with the term "organism" is a part of an ongoing debate within the biological sciences over the necessity for and use of the term at all. Back in the 1980s the paleontologist and evolutionary biologist Stephen Jay Gould felt it was centrally important and even promoted the "restoration" of the term "organism" in evolutionary theory. But there are other words already in use to refer to variants of the organism concept, perhaps an indirect sign that we're grasping. These other labels depend on how independent or interrelated something is. You can find at least a dozen of these options in the scientific literature, from bionts and genets to morphonts, ramets, semaphoronts, and superorganisms. By this chapter's end we'll also explore mathematical, algorithmic definitions of organisms that may actually resolve our confusion once and for all.

Eliminating that confusion is key to better explaining our relationship to the dataome. Are we and the dataome an organism, or are we and the dataome each an organism, or are we something else altogether, drawn from the above list? Because the word "organism" immediately conjures up a mental picture involving a variety of boundaries, both physical (like cell membranes and skin) and conceptual (the idea of something encapsulated in itself, of individuality), and perhaps because of our own prideful sense of identity, we could be missing the right way to look at all of this.

## Enter the Holobiont

Today, in our interconnected human world, many have observed that our essential nature is shifting. We talk about our "hive mind" when ideas are exchanged online, across the globe. Or we talk about being a connected superorganism, like eusocial ants or slime molds. We see ourselves outsourcing parts of our cognitive operation to Google searches in lieu of actually remembering facts and figures, and the rate at which news and ideas spread has accelerated far beyond anything possible before. At the same time our machines are becoming more and more sophisticated, interpreting our actions, predicting our behavior, and amplifying our abilities—from text messages to fly-by-wire planes and cars.

With a little soothsaying, various thinkers and writers have even suggested that we're experiencing a transcendence. In the language of the philosopher Immanuel Kant in 1781, transcendence is about being beyond the limits of all possible experience and knowledge. Exactly how you know that's happening seems a little tricky to me if it's beyond knowledge, but it's an alluring concept. There is little doubt that something is going on with our species, but whether it's truly unexpected or beyond the limits of experience is an interesting riddle. It's also a riddle that the concept of the dataome could help us solve.

Part of the answer may come from extending our concept of organisms and our concept of what it is that natural selection really acts on. The picture of selfish genes pulls us in one direction, saying that selection ultimately acts on genes and the information packages they represent. The picture of group or multilevel selection pulls us in another direction, saying that the units of selection are many, and more interrelated. But what about humans and their dataome?

In 2012 a provocative research paper showed up in the pages of *The*

*Quarterly Review of Biology*, written by the researchers Scott Gilbert, Jan Sapp, and Alfred Tauber. This article proposed a major realignment of thought on the nature of biological individuals among complex, multicellular living things. Central to its arguments is the astonishingly complex and essential array of symbiotic relationships between all large, multicellular life and populations of single-celled microbes—the microbiomes that help make us all tick.

Digestive systems, metabolisms, immune systems, and biochemistry in complex-celled, lifelike plants and animals appear inextricably intertwined with the workings of innumerable species of bacteria and archaea. We, as the Gilbert paper asserts, have never been individuals. That means that the selection pressures on us—our fitness, our Darwinian trajectory through time—have as much to do with the fitness of our microbiomes as anything to do with our own DNA.

That implies several things. The first is that we need to talk about complex life as new kinds of individuals—as *holobionts*. That's the sum total of a host and all of its microbes, and the sum total of all of those genes is therefore a *hologenome*. This also means that a host and its microbes is not the same as a superorganism. A superorganism is a term used to talk about integrated social units made up of the same species, like a termite colony. By contrast a holobiont is made of different species, different genomes, but all acting in ways such that a host's phenotype—the way its DNA manifests in the world—is profoundly affected by its microbial communities.

But that leads to a perplexing question. Different pieces of a holobiont might not always agree on what is best. As a purely hypothetical thought experiment, one group of microbe passengers might be best off with certain foodstuffs coming through their host's system. Maybe pizza and donuts provide their ideal nutrition. But for the host that diet will lead to obesity and impaired cardiovascular function. At the same time, both host and microbes are symbiotically linked—they need each other. Does this mean that they should cooperate or compete?

That tension can go even deeper. When humans are born we seem to accumulate our microbiome both from our parents and from the environment. The lineages of our microbes are to a large extent separate from our own. In evolutionary terms they must therefore also have separate interests. That means that traditional evolutionary theory should treat hosts and microbes (and lineages of microbes) as separate units of selection. But that approach would then backtrack on the appealing unifying power of thinking about holobionts and their hologenomes.

This is a contentious issue in the field of evolutionary biology. It's not clear that holobionts can be analyzed as units of evolutionary selection. Yet if they can't, this means we're still searching around for, in effect, what matters most in natural selection for complex life. It may be that we just have to deal with these complications, studying the coming and going of microbes (they don't always spend all their time with the host) and tracking all lineages and their interactions. Some scientists make this argument, pointing out that hosts and microbes are still better considered as individuals existing within an *ecosystem* than as a holobiont.

But there's another twist to this, a different way of looking at things altogether. In 2017 two researchers, W. Ford Doolittle and Austin Booth, both at Dalhousie University in Halifax, Canada, published an article with a title that began with "It's the song, not the singer." That's not the usual fare for papers on the philosophy of evolutionary biology, and what Doolittle and Booth propose is both intriguing and contentious.

They address the evolutionary tension of multiple microbial and host lineages by pointing out that the net biochemical contributions of microbes to their host change comparatively slowly across generations. You and your offspring do not have radically different microbiome behavioral properties. That suggests that it might be the *patterns* of interaction between host and microbial species—in metabolism or development—that are actually the units of evolutionary selection.

This takes a bit of mental gymnastics to fully appreciate. But the analogy of a song and singer works pretty well. We've all seen real-life versions of this over the years. Imagine a group of musicians gets together to form a band. After some initial struggles they end up producing one hit song—the kind that gets endless airtime. This is the pinnacle of the group's success, and after a happy few years they descend into acrimonious disputes and break up. The song, though . . . well, that just keeps on going, in endless cover versions and commercial soundtracks. In evolutionary terms the song had high fitness: it was selected for, and that drove the (temporary) cohesion of the band.

In biology there may not be any actual songs per se, but there are structured phenomena, such as metabolic processes or patterns of behavior, that are direct consequences of the host-microbe symbiosis—of the holobiont. If those patterns are successful or unsuccessful, they will result in evolutionary persistence or pruning. Within the holobiont there could be all manner of conflicting interests, mutualism or antagonism, but in the end that doesn't matter. What matters is the fitness of the overarching pattern that emerges.

Taking this further, it's tempting to ask whether the concept of a holobiont is also a way to describe humans together with their dataome. Two species engaged in symbiosis, where the song of their interaction is the unit of Darwinian selection.

From the start of this book I've laid out a trail of evidence and proposals to illustrate how integrated our existence is with our external data. And now we're noticing some extremely seductive parallels to the symbiosis between complex, multicellular life and microbial life. Each of us carries our own piece of the dataome. We are inoculated with information from the moment we come into the world. Our neurons carry this information with us. But we also exchange information constantly—not just with each other, but through and with the infrastructure of language, writing, and data storage. This information is not just composed of memes, either. Memes are a data species (not taking that term too literally), rather than the overarching kingdom.

And if we and our data form a holobiont, it could be that the resulting patterns of interaction—across our minds, bodies, and untold variants of informational units—are the things that gauge evolutionary fitness. In the previous chapter I discussed evidence for how the literate human brain is different from the illiterate human brain. Is literacy a selectable trait for our bio-data holobiont? While our dataome pulls us toward ever-increasing energy use, we would still prefer to slow the changes wrought on our planet. Is this an example of the conflicting interests of the members of a holobiont?

At this time, I don't think we can make a definitive statement. But concepts like the holobiont and the ongoing quest in biology to understand how selective processes really play out are critically important. Not only do they speak to the fundamental nature of the phenomenon we call life, they might also provide insight to why humans, and all other species, create all the things that they do.

One of the most remarkable properties of living creatures is to produce novelty in the world—to grow, build, and express the unexpected. Like life itself, that novelty can also stick around, if it's successful at improving the odds of its own future.

## Core Algorithms

I started this chapter talking about the edges of scale for planetary life, and the core metabolic processes that Earth's species collectively exploit. We think that the genes describing the central ways to carry out these types of metabolism (the molecular machines that living things need to use oxygen, or methane, or for fermenting compounds) have been inherited, shared, and swapped across species throughout life's history. That's why we find versions of those genes wherever we look.

But metabolism is a trait, a phenotype, and there are other kinds of phenotypes that we see shared across multitudes of species that long ago separated in terms of relatedness.

There can be similar phenotypes that are actually the result of quite different genes. For instance, pigmentation properties of organisms, from humans and mice to birds and insects, often seem to be driven by similar selection pressures. In some cases, a lightly colored coat in a lightly colored environment makes you less prone to being picked off by predators. But there are different variants (alleles) of the same gene (such as the so-called agouti gene, or ASIP gene) that confer similar coloration, and there are entirely different genes that confer similar coloration (such as the melanocortin 1 receptor, or MC1R gene).

This example of pigmentation is a form of convergent evolution: the independent discovery (for want of a better term) of identical strategies, body plans, and solutions to problems by different genes or by species with widely separated histories. Other famous examples include the camera-eye optics that we find in both vertebrates and cephalopods, and the capacity for atmospheric flight seen across vertebrates and invertebrates. In each case there are hundreds of millions of years of separation of these species from a common ancestor, and different genes at play.

Our usual explanation is that these are simply successful strategies across a range of environmental conditions. It is therefore to be expected that life's ceaseless experimentation will, sometimes, land on near identical outcomes, even if the underlying genetic history is different. Evolution is a powerful search engine that converges on the same answers, searching from the bottom up.

But, like the concept of the holobiont, it could also suggest that the "song" is the unit of evolutionary selection—the song being the algorithm that chooses advantageous pigmentation, or that encodes the optical physics of camera eyes, or that directs the essentials of aerodynamics and heavier-than-air flight. Or indeed the algorithms for energy utilization that are represented as metabolic processes.

Perhaps there are situations where the most efficacious conceptual (and computational) framework for understanding life is one that embraces the upper layers of an informational biosphere. If genes are merely representations of information that happen to describe a way for itself to replicate and spread, couldn't something like these songs of convergence be the same?

The spread of flight among species can be conceptualized as a spread of the song of flight, the algorithmic invention. For convenience I'll call it a core algorithm, or "corg" for short. Sometime over 320 million years ago, within the population of insects, the corg for flight emerged on Earth. And, just as for genes and groups and holobionts, what matters for Darwinian selection is the fitness of the information pulling the strings, the thing behind the curtain of the great and powerful Oz. Or to put this another way, what I'm suggesting with corgs as conceptual tools is not so different from the notion of genes as manifestations of self-propagating information. It's just that the corgs' influence may flow in a different direction, an important characteristic that we'll return to.

For the flight corg, the informational algorithm is held in no single place, but rather across all the genes and species that enable it in small and large ways. The fitness of the corg, the selection pressure, is measured from the fact that flight enables the growth of different branches of life on Earth. It allows the total biosphere to expand into new niches. Therefore, there has been growth and persistence of the flight corg, as well as everything supporting it.

The same rationale applies to a corg for the general phenomenon of metabolism: the algorithm is the process of energy flow and utilization itself rather than the individual mechanisms by which that happens. Or for the phenomenon of camouflage (rather than merely pigmentation): as we've seen, subterfuge or deception must be very ancient, and the corg for that enables the growth of the whole living system on Earth, and therefore the growth in number of its own implementations.

The example of flight is a particularly good one for other reasons too. It offers an insight into how corgs and dataomes might connect, as simply different sides of the same many-faced die. Humans are biologically incapable of flying. We have nonetheless invented ways to get ourselves into the air. Some of those ways are less obviously inspired by what we observe in the world, such as the balloon or airship. Others are very direct, such as the forms of wings and the idea of propulsion by pushing at or blowing air around you, or simply gliding and soaring in the currents that flow through the atmosphere.

It's certainly plausible that we would have come up with ways to fly if there were no other biological entities that used the atmosphere this way for us to learn from. We might still have chucked thin bits of wood off a cliff, or watched materials be blown into the air by wind. But we could also imagine that the flight corgs that emerged and were manifested long ago in other species to enable flight have directly exerted their influence on us. The difference is that we don't seem to have subsequently encoded flying traits into our DNA. Instead we have encoded them into our dataome. The corgs for flight have found a way to aid their propagation by manifesting themselves in our data, and through our data into us and into actual flying machines.

While there are individuals among us who can probably build a basic airplane just using the knowledge in their brains, none of us could construct an entire Boeing 787 or Airbus 350, much less a Saturn V rocket or one of SpaceX's Falcon Heavy lifters. These extremely complex flying machines are entirely a product of cooperative behavior. They are also impossible without a sophisticated dataome to store, manipulate, and study reams of data and information, from aerodynamic theory to simulation outcomes, material design, and wiring diagrams. And the amazing thing (if you buy into this) is that the flight corg is managing to instantiate itself (become embodied) into both biological and non-biological forms, and it is able to do this thanks to the dataome.

Now, I know that if you're full of caffeine you'll be itching to say that everything is looking a lot like a hierarchy of corgs. Because isn't the dataome itself representing a corg? And the whole system of life on Earth is also representing a corg. So, if there are corgs within corgs, which corg is most important?

I don't know the answers to these questions. It may be that none of this represents an accurate, all-encompassing model for life on Earth. Perhaps, in the end, it does all just boil down to the informational algorithms embodied in genes, along with a lot of messy complications. The final arbiter—that could in theory yield a quantitative answer—will have to be statistical modeling: a composite of everything summed up according to each of the competing theories, correlated and judged at the altar of propagation odds and Darwin's distant gaze.

Nevertheless, I find the concept of core algorithms pretty compelling, especially because there are entirely independent lines of thinking that, quite unexpectedly, seem to be talking about very similar phenomena. One of those comes from philosophy, one comes from pure mathematics, and another comes from physics (and we'll come to that later). The philosophical line of thinking comes from a surprisingly practical method of study mainly credited to Edmund Husserl and his student Martin Heidegger called *phenomenology*. It deals with objects and events as understood in human consciousness. That doesn't sound very promising in the present context, but the application of phenomenology, particularly by the mathematically trained MIT philosophy professor Gian-Carlo Rota in the late twentieth century, yields some intriguing stuff.

Rota homed in on the notion of abstract ideas or structures being instantiated in the world. In one exposition he used the example of an airline timetable. Such a timetable might be printed, it might momentarily exist in your brain or it might be in a computer database, and it might exist in different versions, updated or adjusted. But in that case what really is an airline timetable? It is something that exists in a more abstract layer of reality. Another example is a key. This can be a metal object, but it can also be a plastic card or a code that you enter on a computer. The key's "keyness" exists in a way that is external to its physical instantiations, although entirely dependent on the existence of a hierarchy of things, like doors, locks, security, and so on.

The mathematical link stems from the early twentieth century, when mathematicians in the 1930s and 1940s conceived and produced what may be one of the most abstract and wacky-sounding mathematical formalisms (at least to any non-mathematician). *Category theory* is a theory of other theories, or, more accurately, a way of abstractly representing mathematical fields so that they can be related to each other. Topology and algebra might seem like they're two entirely different things. But category theory lets you talk about the *structures* or *patterns* that are common to both. It lets you describe, for instance, how an algebraic structure (like a method for solving an equation) maps onto a topological structure (like how geometry works on a curved surface). That's quite awesomely abstract, but it's very similar to how many modern computer languages utilize structure templates that can be endlessly customized to fit tasks that are superficially different but structurally the same. Like how ordering a box of snacks online is actually structurally like going into a store and taking something off a shelf and paying for it.

Phenomenology and category theory both grapple with the notion of moving to a higher level of abstraction in order to better understand how the world (and reality itself) really works when it's full of imperfect things like us. They seem to be expressing very similar conceptual strategies to what I've described for understanding life and information in terms of core algorithms.

It's important to be clear though that an information-centric description of life on Earth is not meant to suggest that phenomena like corgs (or indeed genes) have been emplaced or have predated their implementation in the world. To say that corgs came from elsewhere, outside of the world, would be a type of pathetic fallacy (of assigning intent or human qualities to nature) taken to the idiotic extremes of creationism or intelligent design. No. Instead, the corgs and their information structures are just as much a part of the extraordinary properties of Darwinian selection and evolution as the bits and pieces of the physical world that they are manifested in. They have to form and attain focus as much as the humblest blob of a primordial lifeform. Once they do, they are hard to get rid of. The universe is spectacular because it is an engine of invention that starts with a near formless soup of primordial matter and energy and builds structure and complexity that persists. And that story, about where anything new ever comes from in the first place, is what comes next.

## Open Dreams

For millennia human philosophers and poets have mused on the origins of ideas, some with more lyrical flourish and sense of wonder than others. In the 1700s the Scottish philosopher David Hume wrote about how the mind consists of perceptions, divided between impressions and ideas. He thought that perhaps our imaginations, our little idea factories, were not boundless but rather driven by the continual combination and recombination of a finite set of impressions and ideas prompted solely by sensory inputs. This was a bit bleak, but also admirably rational, leaving little room for what we might today call pseudoscience or mysticism.

Others, like the German polymath Gottfried Wilhelm Leibniz in the 1600s, also felt that all human ideas and imagination came from the assembly of more basic, fundamental, immutable concepts. But Leibniz was way more exuberant. He was fascinated by the notion of using symbolic representations of these fundamental concepts to produce new ideas by simply exploring their combinatorics. A little ironically, he wasn't quite treading original ground. Going back to the 1200s, in Spain, a Jewish mystic called Abraham Abulafia had experimented with combining letters of the Hebrew alphabet in quasi-random ways. He claimed that there were secret rules for this "science of the combination of letters" that, basically, came right down the pipe from God.

If this sounds like Claude Shannon's scientific experiments with the rules and probabilistic nature of language, that's because it really does overlap. The assembly of ideas by combining symbols is connected to a very modern area of research: NLP, *natural language processing.* NLP is a very hot and challenging area of current machine learning and information science efforts because it's about producing systems that can both interpret and write natural-sounding text—systems that capture what humans do.

But for Leibniz, who knew about Abulafia's earlier experiments of letter combinations, natural language wasn't the holy grail. Instead, he felt it might be possible to build a machine to actually generate new ideas, a machine capable of answering all of our deepest philosophical questions. He called this his "great instrument of reason," and it would use both his proposal for a "calculus ratiocinator" (a universal logical framework) and his "characteristica universalis," a universal language he envisaged as encompassing all mathematical, scientific, and metaphysical concepts—a language of human thought. Whatever Leibniz used to drink for inspiration, it was good stuff.

The catch with these fantastic proposals was that for most other thinkers and observers at the time, as well as today, it felt like something was missing. While the properties and structures of the universe seemed to be expressed mostly in mechanistic, endlessly predictable patterns, they could also misbehave. Indeed, by the late 1800s, with work from people like the French mathematical genius Henri Poincaré, it was clear that there were chaotic, fundamentally unpredictable behaviors embedded in the mathematical bedrock of the world. There are equations of physics whose internal feedbacks result in exquisitely sensitive and unruly outcomes, where the tiniest numerical changes are amplified to generate radically different solutions. From Earth's weather to planetary orbits, the cosmos isn't really clockwork, it just looks that way. It wasn't obvious that a machine or clever symbolic language could ever capture the nuances of all thought or the complexities of nature, much less produce novel, imaginative ideas by just turning a handle.

That remains the situation today. One of the most rousing modern examinations of this problem was an essay written in 2017 by the computer scientists Ken Stanley, Joel Lehman, and Lisa Soros, titled "Open-endedness: The last grand challenge you've never heard of." The basic gist is that when we examine the nature of life and biological intelligence, it is very clear that these phenomena are incontrovertibly epic in terms of complexity and novelty. Although the universe started out comparatively simple in its structures, nature has "invented" things like Darwin's awesomely entangled bank out of thin air. Or rather, not thin air, but billions of years of experimentation, constantly layering new stuff on top of previously new stuff.

Stanley, Lehman, and Soros couch this in computational terms: Evolution on Earth is like a single run of a single algorithm that invented all of nature, akin to what Richard Dawkins has called an "information bomb" and what I've suggested as a master corg for the planet. And that algorithm doesn't seem to stop. It just keeps on going, even after billions of years. They call it the "never-ending algorithm" that keeps spewing out greater complexity, a process that we can also call open-endedness.

The reason that open-endedness is a grand challenge is that for all the machines and software we've built, for all of our searches for the near-mythical possibility of generalized artificial intelligence, we seem to make no major progress toward developing anything that exhibits life's endlessly creative open-endedness. Computer scientists and some molecular biologists are certainly desperately keen to find such algorithms or chemical milieus: systems that generate actual ideas or totally unexpected, functional code or useful outputs. Or molecular systems that inch toward the long-sought-after magic of the origins of life. But in all cases these algorithms or chemistry sets keep grinding to a halt.

There are classes of computer algorithms called Evolutionary Algorithms, or EAs, that are aimed in part at solving specific mathematical or statistical analysis problems. EAs can use approaches inspired by gene-driven biological selection. They might generate a population of sub-algorithms that are then selected—as in actual natural selection—for their fitness, their ability to get the closest to an answer. These selected sub-algorithms are then the parents of the next set. That subsequent generation is therefore a population of algorithms descended only from the more successful ancestors.

These are superbly powerful computational tools, often homing in on answers to problems that are nigh-on intractable using conventional means. But even EAs always seem to get stuck at some point. They stop inventing. They don't endlessly generate novelty.

Many researchers over the last several decades have worked to break the impasse, looking for tricks to help codes and machines somehow take a leap of imagination, for want of a better word. There have been a lot of incremental successes, especially in combination with current deep-learning algorithmic strategies. But nothing thus far seems quite capable of not just solving problems but actually *creating* the next set of problems to solve. In other words, doing what biological evolution, and minds, have been doing all along on Earth.

Stanley and his colleagues use a cute example to describe this: Trees were once a novel solution by plant life to the challenge of survival and reproduction in a world of problematic, hungry animals and competing plants. But that also, eventually, created the opportunity for giraffes to emerge—for animal life to hit on another novel solution to survival, the solution of long necks. True open-endedness is what creates opportunities for future open-endedness.

What is it about nature that lets its algorithms turn the dial up to eleven, while we're still fumbling around somewhere close to perhaps one or two? The suspicion is that a part of the magic of open-endedness comes from the coevolution of entities: the back-and-forth of one entity creating a new condition of some sort for the other, which in turns creates a new condition for the first, like the trees and the giraffes. There are also similarities in this to the generative adversarial networks we encountered earlier. Ken Stanley and his coauthors point to the phenomenon of social organization in humans that allows our inventions to build on past inventions over very long periods. We didn't forget about wheels once we invented them, we just keep adding layers.

Ideas that can propagate back and forth through the network of human minds can trigger new ideas in a cascade that amplifies change and novelty beyond anything that even a single human brain's 86 billion neurons are capable of. To me that surely means that the dataome is an integral part of true open-endedness in human intelligence. Nothing else serves such a role for capturing knowledge and enabling complex interactions with that knowledge for billions of humans, whether sequentially or simultaneously.

Another suspicion about the requirements for enabling open-endedness in the world comes from studying the behavior of non-equilibrium thermodynamics. We briefly encountered this in seeing the effort required to build the low-entropy devices of computers and the dataome. Statistical mechanics may capture the typical, averaged-out properties of big, many-pieced systems—from gases to solids to populations of organisms. But there can still be fluctuations, temporary deviations from equilibrium, and temporary violations of the rules, creating novelty.

In 1994 the mathematician Denis Evans and the chemist Debra Searles came up with a mathematical proof of the "Fluctuation Theorem," which says that drops in entropy (sometimes, and a bit misleadingly, labeled as negative entropy) in large systems absolutely can happen. It's just that the bigger those drops are, the less and less likely they are to persist. When they do persist, they can correspond to some of the most complex, hard-to-quantify patterns and behaviors that we encounter in nature; from the turbulent swirls and structures in liquids and gases to the shapes of enzymes and the structural roughness of a crinkled-up DNA molecule.

There could be one further missing ingredient when we try to construct algorithms to produce open-endedness, or to capture the algorithmic nature of living systems. To recognize that missing ingredient, it helps to imagine that you could whisk away all life from Earth, scrub it and sterilize it. You would still have an exquisitely complex planet: a world of random mineralogical variation and geophysical jumbling, all coated in the multi-scale chaos and flow of oceans and moist atmosphere, every piece still engaged in its own chemical cycles and fluxes of energy and physical interaction. There would still be endless day-night transitions, seasonal changes, and over tens of thousands of years there would still be orbital oscillations that bring the planet into fundamentally different climate states, and then out again.

If you re-inoculated this sterile planet with single-celled life it's not hard to imagine that once that life started interacting with the complexities of its environment, things are going to get interesting all over again. Earth itself represents an incubator of open-endedness in the universe. But when we construct an algorithm or laboratory experiment to try to "create" the innovation of open-endedness we're not really exposing it to that incubator. Instead it exists in a simplified virtual environment, in a sterile test-tube, or within a limited set of inputs and parameters.

It probably takes an entire planet to make life like us. Not a restricted corner or limited ecosystem, but the whole shebang of a messy, violent, changeable sphere of minerals and energy churning and changing across four and a half billion years. This is what builds (and is built on) core algorithms, and what keeps on making novelty. For humans, and human minds, the dataome is a key piece of that very environment. As much as it started somewhere with us, it has also coevolved with us, constantly amplifying and helping our minds attain open-endedness.

## Organism (Part II)

I've discussed some of the problems we face in categorizing nature, when nature doesn't mesh neatly with our symbolic languages or cognitive tendencies. The interesting thing is that the more we investigate the world through the lens of information and mathematics, the more we see ways to capture the characteristics of the world that sidestep the pitfalls of language and our minds. That pesky label, "organism," grasps at a sense of individuality, and it turns out that perhaps we can hand that concept over to algebraic, or algorithmic, analysis. That we might do this offers hope that there will be ways to further disentangle the head-throbbing complexity of life, evolution, the dataome, and core algorithms as I've described them.

The evolutionary biologist David Krakauer, whose work on organism selection we encountered earlier, together with several colleagues, has looked at the notion of an information theory of individuality. To do this the researchers dip into Claude Shannon's bag of tricks, using information entropy and mutual information to try to describe biological individuals in terms of their information contents and their use of information. You'll recall that mutual information is the degree to which one entity contains information about another. In a system of communicating parties that's like you and a good friend on a phone call. When they extend this concept to treat a biological entity and its communication with its environment (all the bits and pieces of the world, including itself), ways to define individuality seem to emerge.

The details are quite technical, but the idea is that we can think about a living thing as a system of associated processes. If we (or indeed the organism itself) have knowledge of those processes we can predict what the organism will do, or should do in the future. But not all of these processes really help with that prediction, especially those further and further removed from the organism, with less and less mutual information. This suggests that a way to define an individual is to add up all of these processes (things like the motor system, the metabolic system, and so on) until adding more makes no difference to the ability to predict its future behavior. At that point, in the sum total of those processes, we will have identified the entity that can be treated as a true individual.

In some situations the informational boundary of an individual could easily encompass many physical structures that we might otherwise label as separate organisms. This could be true for a bacterial colony, a coral reef, or a multicellular entity like a human. Computing the predictive power of the information, its surprise and meaning, can yield a quantitative statement, a way to decide where an individual (or organism) really ends. Perhaps we will be able to apply these tools to questions such as whether or not animal-microbe holobionts are valid entities from the point of view of evolution. Perhaps too we will be able to answer whether or not humans and their data are a holobiont—inextricably bound on their journey through evolutionary time.

# 6

# THE INFORMATION RIVER

*Life and death appeared to me ideal bounds, which I should first break through, and pour a torrent of light into our dark world.*

—Mary Wollstonecraft Shelley, *Frankenstein*, 1818

If the rules of life are built around information that propagates itself, it stands to reason that we'd want to figure out exactly how (and, eventually, why) this happens. What are the rules of information that lead to the rules of life? And how, in detail, does information exert itself on the physical world? The answers aren't just of academic interest. They'll surely govern our options in the face of all the intriguing and dangerous things going on for our species, from technology that replaces our cognitive skills to the climate changes rattling our environment.

In a very real sense all of us, and all of what surrounds us, are pieces inside an ongoing four-billion-year-old game. It's not a game with a perfectly prescribed set of rules. Instead it's a game governed by rules that emerge from the game itself. One might say that the very object of the game is to produce its own rules. The ultimate players of the game are also hard to spot. They are even more ancient, and have no discernable physical form—they are the root properties and predispositions of the universe; they are everywhere, yet nowhere in particular.

In your immediate vicinity the game takes untold quadrillions of tiny, similar objects (we call them atoms) and assembles them into more-complex structures that, as they grow, can become less and less like each other. We call these structures molecules, organelles, cells, individuals, and species. The more complex these arrangements are, the more information they represent and propagate. That information governs physical structures and the endless back-and-forth of atoms and electrons, but it also consists of the rules by which those things are judged for survival.

If this all sounds pretty bizarre, well, it is. Our daily experience of the world doesn't really look anything like this. If I throw a box of LEGO bricks onto the floor they always disappoint by just sitting there, not doing anything. If I empty a bag of cement onto my yard a nice patio feature doesn't assemble itself. But that's merely a fault of timescale and our human perceptions. Wait for a million years and chances are that the atoms of now-dissolved LEGO blocks and cement will have indeed done something new—albeit not what we might have wished for. The game is still afoot.

The assembly and disassembly of structures also involves energy transformation and *flow*. Two atoms can configure themselves in a joined state that has a lower net energy than when the atoms are separate. It's like corralling two enlivened toddlers and making them hold each other's hands for a modicum of peace. In some cases, each atom can contribute and share one of its electrons with the other atom. The net result is that the atoms are bonded (a covalent bond in chemical terms) through the association of this pair of electrons—one from each parent atom. It's a calming and favored atomic hand-holding because the energy of this joint system is lowered, as each electron finds itself less constrained, with the luxury of shared space to spread across. The lost energy produces molecular vibrations, sometimes radiating away as photons of light or being passed along to other jostling molecules, and this type of bonding can be even stronger when four or six electrons end up in this kind of polyamorous arrangement.

A conceptually simpler form of chemical bond is an ionic bond, when, for example, one atom fully donates an electron to another atom, causing both to become electrically charged, one positive, one negative. These ions are then attracted to each other, and to a state of lower net energy. You've eaten lots of ionic bonds during your life—this is what makes sodium chloride, or sea salt. But it's not just energy that moves around; the informational content and capacity of these tiny components also evolves. Imagine two bonded carbon atoms. Individually, in their electrically neutral, fully relaxed state, the only data they can readily "store" is in their internal magnetic properties—a so-called magnetic moment. But if they're bonded to each other, the molecule they form can rotate and vibrate (like objects connected via a spring, or two twirling toddlers). There are suddenly more degrees of freedom, more options for the behavior of this two-atom system, more surprise, and more information storage.

A molecule of three atoms, with two bonds, creates even more outcomes. The technical terms for its modes of molecular vibration are: symmetrical stretching (think push-ups), asymmetrical stretching (like a boxer's arms pumping back and forth), scissoring or bending, rocking, wagging, and twisting. And because of quantum mechanics these behaviors have discrete, quantized states, like a set of labeled positionings on a rotary dial. The potential for even the simplest molecule to hold information increases with every additional atom.

Even without accounting for these gymnastic moves, with 90 naturally occurring elements on Earth the number of basic combinations and structural permutations for molecules is staggering. Take a relatively boring molecule with the chemical formula $C_6H_6$—that's six carbon, and six hydrogen atoms. Boring or not, there are 217 potential structural forms or isomers of this molecule that obey the rules of chemical bonding and stability. That's 217 different ways to hook these twelve atoms together.

Go to a slightly different chemical formula like $C_7H_4O$ (with an oxygen atom) and this leaps to 11,332 structural forms. Add in new elements and the numbers climb exponentially.

Each molecule is a unique configuration, an isomer, with its own chemical, vibrational, and rotational properties. The interaction of any of these isomers (if they happen to exist in nature) with the environment of other molecules will produce new structures, new energy exchanges and transformations. Consequently, one simple chemical formula represents a colossal potential for representing information and for the interaction of information with other information.

In this way energy flow and information flow are intimately connected. Energy flow drives information flow; information flow apparently drives energy flow. Consequently the fundamental molecular actions that are the bedrock of biochemistry and living systems inevitably shift information around. That is not to say that there is information that means anything to us as humans. You are unlikely to find shopping lists or gossip within the metabolic reactions of an amoeba. But there is most definitely information; an uninterpreted imprint of molecular properties and histories.

Most critically of all, there is the ebb and flow of rules. Every newly formed molecular structure unlocks a web of potential future interactions, a multitude of new possibilities that may never have existed before. As with evolvability in biological species, the formation of some molecules unlocks greater potential variety in future molecules. Those variations will also come with attached *probabilities,* since not all chemical bonds or structural forms have the same likelihood of occurring. In other words, there are new rules being spontaneously, and continually generated at the deepest level of atoms and molecules.

But what happens to the rules as more and more complex interactions and structures have a chance of occurring? As populations of structures interact among themselves and with other populations and the physical environment? These questions are also essential to understanding the origins of life and the emergence of living, evolving, novelty-rich systems, and we don't have any final answers yet. We do not know what the transition from not-alive to alive really looks like, or how that happened on Earth, or indeed might have happened anywhere else in the universe. But we do have clues.

## Talking Things Over

Picture a weird 1950s pulp-novel version of an alien creature. It's got an oversized elongated icosahedron head (a three-dimensional object with twenty sides), a ribbed vertical tubular body, and spindly spider-legs running around its lower end, letting it perch on a surface like a carefully balanced drilling rig with a nasty looking drill bit. It's ready to invade Midwestern America and start attacking plucky, strangely over-competent teenagers.

Now shrink this object down to a couple of hundred nanometers in size and you have something that is not an alien, but is instead a very real, very successful entity at the hairy border between what we think of as life and non-life. This is a bacteriophage, a tiny molecular machine.

The viruses known as bacteriophages are among the best studied viral strains in science because of what they do to bacteria that can be pathogenic to humans. The way bacteriophages work is that they need to infect a bacterium in order to reproduce. Two ways this can happen are through a process called the lytic cycle or another called the lysogenic cycle. In the first the virus drills inside the bacterial cell, hijacks the bacterium's biochemistry to make many copies of itself, and then (brace yourself) escapes the bacterium by making it explode. The phages actually do this by expressing genes at just the right time to generate proteins that poke holes in the bacterium's cell wall, allowing external water to flood into the cell and burst it.

The lysogenic cycle is at first a little less dramatic, but quite insidious. In this case the little drilling rig of a virus inserts its genetic code (which may consist of anywhere from just a few genes to a few hundred) into the bacterial genetic material itself. When the bacterium reproduces by cell division it then dutifully also copies the phage genome, and passes it on. At some point the genetic material of the phage can extricate itself from the bacterial DNA and start making new phages—essentially switching back to the cell-exploding lytic cycle again.

But how do the viruses know which infection strategy to pursue and when to switch modes? It turns out that one of the factors correlating with the switch is the total number of bacteriophages present at a given time—a factor that is perhaps related to maintaining a workable balance between a population of viable hosts and a population of active phages. The question then becomes, how do the viruses keep count? Recent research is beginning to reveal an extraordinary answer for these superficially primitive entities: they talk to each other.

When phages infect a bacterium and kick off the cell-exploding lytic cycle, they also release a chemical signal. In one strain of phages this signal is in the form of a specific tiny compound, a peptide molecule made up of just six amino acids. In an environment where there are many phages doing their infectious thing simultaneously, the chorus of peptides grows louder and louder. When it reaches a critical threshold—a certain level of peptide concentration—the viruses then switch to their more low-key lysogenic cycle, in effect lying dormant rather than destroying all of their hosts. In other words, these tiny entities, consisting of nothing more than a few hundred thousand atoms, exchange information. And that information exchange is critical for maintaining their propagation.

In 2017, shortly after this phenomenon was first reported, four independent research groups set out to try to discover how these signaling peptides, now called *arbitrium peptides*, are made, what variants there are, and how the phages pick them up and respond. So far, at least fifteen types of phages infecting soil bacteria have been found to produce arbitrium peptides of different flavors. Each strain of phage seems to talk with its own bespoke peptide molecules, meaning that only closely related viruses will be receptive to the messages.

This alone would be impressive enough sophistication for such tiny entities, but things get even more interesting. It seems that when the arbitrium peptides are sensed by a phage via a molecular receptor (a molecular-scale lock-like mechanism that reacts to specific peptides), the receptor doesn't just pass on the message to the reproductive genes in the phage, it also interacts with other parts of the viral and bacterial genetic code. There's a possibility that it can trigger interference with the bacterium's own genetic activity—prepping the victim for what comes next. The viruses have evolved a very targeted engagement with the information in the bacterial genome.

There is also evidence of a complex game afoot between the viruses themselves. If the peptide concentration reaches the switch threshold, there could actually be incentive (in the broadest, non-cognitive sense of the word) for individual phages to cheat and ignore the shutdown message. If their fellow phages all lay down their arms, the cheaters can gain an advantage by continuing to power on, replicating merrily away without competition.

The precise behavior likely varies in different circumstances and between different viral strains. But a study done in 1999 by the scientists Paul Turner and Lin Chao modeled the behavior of phages in this kind of situation as a form of the famous Prisoner's Dilemma—a core scenario in the field of game theory. The Dilemma pits decision makers against each other, so that cooperative choices offer modest benefits. But cheating, or not cooperating, can confer a significant benefit—or cost, depending on what the other participants do. In the cases of phages, the cooperation or defection of the viruses seems plausibly related to the evolutionary fitness of the strains; in some cases selfishness wins out for a fitter strain, in others cooperation prevails among strains that are more equally matched.

Bacteriophages are not the only talkers. There are plenty more examples of viral communication. The polio-causing virus comes in distinct strains, but these seem to be able to clump together when necessary and actually exchange genetic material in a way that increases their cell-killing ability. Findings like these have given rise to a very different picture of the viral world than we previously imagined. Whether we consider viruses as life or as almost-life, they definitely seem to share and sense information. They also seem to be bound by the same kinds of dynamics of cooperation or conflict that extend all the way up to animal species, and to our own behavior, from interpersonal relationships to the stock markets. The nascent field of "sociovirology" is likely to be a hotbed of discovery in coming years.

That cells of microbes or animals communicate with each other is perhaps less shocking, but no less intriguing. Bacteria certainly signal chemically and electrically, as do complex cells. But all types of cells also release what are called extracellular vesicles. These can be bud-like extrusions of pinched-off cell membrane (so-called *ectosomes*) or larger pieces originating from structures deeper inside the cells (*exosomes*). All of these weird cellular children go off into the wider world. The exosomes in particular can carry along all kinds of stuff, from proteins and RNA to bits of DNA and the chemical products of metabolism.

It's thought that part of the purpose for cells squeezing out these structures is basic housekeeping: discarding cellular dysfunction and filth that gets in the way of keeping a cell in balance (perhaps even regulating cellular entropy). But it's also suspected that exosomes play a significant role in intercellular communication. They are literally informational, perhaps functional, parcels for sharing with other cells. That sharing may be problematic and involved in diseases, like cancer or metabolic disorders, or it may be beneficial, enabling cells to trigger or gain immune functions.

The point to drive home here is that at both ends of the range for life, from the simplest viruses to the most complex animals and societies, the ebb and flow of data, of useful information, is without question. That flow is tied to the basic actions of molecules and the transfer and release of energy. It can also seem as if it is in service of evolutionary *goals*. Those goals might be survival or replication, both of individuals and their genes and of populations and species.

Of course, as with the properties of genes, while the concept of goals is useful shorthand, it can't be taken too literally. We witness the existence of genes, viruses, organisms, and species because these are the phenomena that keep on going, that *can* persist in the world. It just so happens that the capacity to predict the future, to anticipate outcomes (entirely mechanically), improves the odds of persistence.

## The Life Thing

But we're getting ahead of ourselves, because even if these processes of information flow are part of the game rules for the phenomenon of life, science is still grappling with just what kind of phenomenon that is in the first place. Back in the 1940s the famous physicist Erwin Schrödinger (known for his essential contributions to a mathematical formulation of quantum mechanics) gave an influential lecture series in Dublin called *What Is Life?* This was before we understood the details of the molecular basis for life (and in fact Schrödinger made key predictions that influenced the eventual decoding of the structure of DNA). For Schrödinger, as for us, a central puzzle was the connection between energy flow and life. For him the question was: How—in a universe governed by the second law of thermodynamics and the relentless growth of entropy—could anything like life crop up at all?

It's one thing to show that life is at least compatible with thermodynamics. If an organism amounts to a patch of lower entropy in the universe, it will be offset by increased entropy in its environment. But it's an entirely different thing to explain how that situation happens in the first place. Schrödinger ended up musing that perhaps there's some kind of new physical law behind life, something we're missing in our picture of reality, a hidden structure.

We're still wondering about that today. There is a suspicion among many scientists that a hidden structure may involve taking the ideas of information theory and, in a sense, shoving them back *into* physics. Claude Shannon's clinical notion of information didn't bother with the *meaning* of that information. However, we're seeing that meaningful information for life has to do with survival, and with the delineation of individuals. Both of those things tell us that when an organism can process information about its environment, it can improve its probability of carrying on into the future—by taking actions predicated on the physics of the world.

This is all pretty logical, but there are complications. One is that if life is just a representation of self-replicating information (like the selfish genes), then does this mean that information grabs onto and uses information? Like some kind of ball of sticky dough? Another complication has to do with life and death. Imagine you are out for a walk and you come across two birds on the path in front of you. One bird is hopping around, very much alive, but the other unfortunate creature is dead, its little feet sticking up from its feathery corpse. Both of these are *life,* but only one is alive. Both have lineages, but only one has the potential for its lineage to continue into the future, and to acquire more functional information about the world. Yet the dead bird is the result of billions of years of life using, generating, and propagating information as much as the living bird.

I don't think any of us would have a problem with calling the dead bird life but not alive because we have prior knowledge telling us how that corpse came to exist. But the intriguing thing is that if we stretch that standard just a little, then shouldn't we call a house, a car, an internet meme, or a collection of Shakespeare's plays life? Equally, what does this make the pieces of autonomous software or backup systems that duplicate data in our dataome? If these phenomena actually straddle the divide between not-alive-life and alive-life, then perhaps they can provide clues for putting information back into physics and figuring the whole messy thing out.

## ALife

In the middle of a chilly January in 2019 I'm sitting in a room at the University of Tokyo, nursing a hot tea and listening to a hand-picked group of scientists talk about some righteously crazy stuff. That discussion includes how alive-life doesn't just involve information, but may really only *be* the flow of information from one place to another. Or how strange things called Maxwell's Demons can erase bits of information. Or what an "information ratchet" is. And how there may be an ultimate limit to any kind of computation that takes place in the universe.

This is, to say the least, a specialized gathering. About 20 researchers have been tasked with hashing out ideas for how to best study the fundamental principles governing living systems. This crowd is mostly drawn from the esoteric, freewheeling, enormously creative, and occasionally disparaged field of Artificial Life, or ALife.

What this field of research even does exactly is, to be honest, not so easy to pin down. A part of the reason for me being here is to try to understand this for myself. I've also been asked by the workshop organizers, from the Earth-Life Science Institute at the Tokyo Institute of Technology (Tokyo Tech), to be the meeting's eyes and ears as an informed outsider. My task, which I am increasingly nervous about, is to try to synthesize and distill the discussions for everyone. To help them sculpt a narrative of accomplishments and critical questions each morning, before getting quite enough of that steaming tea.

To give you a sense of why this is making me nervous, it helps to know the diversity and intellectual oomph of some of the participants. Take Sara Walker, a professor of theoretical physics and astrobiology at Arizona State University. Among her many areas of study is how the flow of information in chemical and biological systems reveals their "aliveness." But this is more than just assigning a label to these systems. Sara's work suggests that major evolutionary changes may correspond to fundamental shifts in that information flow—from bottom-up to top-down in a hierarchy of molecular, organismal, and species relationships. It's an intriguing proposition that we'll come back to momentarily.

Or take Jim Crutchfield, a professor of physics at the University of California, Davis. Back in his graduate student days in the late 1970s Jim was one of a group of young scientists who devised a computational toolkit to help them beat the game of roulette. They famously (or infamously) tried, with some modest success, to take on the casinos in Las Vegas. These days Jim has a guru-like presence, with a knack for working on and explaining extremely esoteric ideas in areas ranging from statistical inference to evolutionary theory and quantum dynamics, to name but a few.

Then there are people like Steen Rasmussen, a physicist and director of the Center for Fundamental Living Technology at the University of Southern Denmark, and a near lifelong member of what he calls the "moving festival of ideas" that is the field of artificial life and artificial intelligence. There's Takashi Ikegami, a colorful and innovative professor of robotics at the University of Tokyo and an expert in swarms—both natural and artificial. And there's Eric Smith, a longstanding member of the Santa Fe Institute in New Mexico and a physicist at Tokyo Tech, considered by many to be one of the greatest living thinkers on the origins of life and life's deepest properties. His own informational processes are exquisitely well-tuned. He's perhaps the only person I know who can speak in complete, perfect, beautifully structured paragraphs without pause.

The workshop, not unlike aspects of the disparate areas of research represented in the room, is turning out to be an unwieldy creature. The field of ALife has had its ups and downs. Some would say that its history goes back to the earliest efforts to build mechanical, clockwork simulacra of life, like the famous Digesting Duck built by Jacques de Vaucanson in

1739—capable of quacking, drinking, and (through some sleight of hand) eating and defecating. But ALife's more scientific, modern incarnation arguably emerged between the 1940s and the 1970s with deep ideas to do with "automata"—machines that obey internal logical rules for making decisions, and for replicating themselves. We'll revisit this when talking about the contributions of John von Neumann, but perhaps the most publicly famous automaton invention was the Game of Life, thought up in 1970 by the mathematician John Conway.

In this game, entities known as *cellular* automata are constructed that perpetuate themselves in an artificial (virtual) two-dimensional universe of cells in a grid. The properties of these cells follow a set of rules that apply in discrete time steps, just like in a turn-based game. Moment by moment the cells change their state—in simple terms you can think of them changing color—depending on what their nearest neighbors are doing. Specific initial conditions (colors and patterns) assigned to the cells can, quite literally, take on what looks like a life of their own.

Some groupings of cells can self-replicate, some can glide through this abstract grid space, others cause seemingly chaotic, unpredictable changes across the grid. It's endlessly fascinating, and has spawned—no pun intended—a vast literature, along with fundamental work on mathematics, complex systems, and computation, most notably by the scientist Stephen Wolfram. The allure of Conway's Game of Life is that simple rules, and information, can give rise to a dynamic stew of entities, computing elements of their futures on the fly.

By the 1980s there was hope that ALife research, whether in software, hardware, or even in wet chemistry, could lead to the development of new kinds of living things. Researchers were finding tantalizing cases and unexpected applications for their studies. There were explorations using versions of cellular automata for replicating patterns in nature, from snowflakes to plants. There were simulations of "boids"—artificially swarming and flocking entities that could sometimes, almost magically, solve problems of finding pathways through obstacles in very natural-looking ways. There were also applications to the study of animal behaviors, and even the nature of evolution, as well as work that would eventually seep into the foundations of what we today call machine learning.

But, a lot like the optimism that existed in the 1980s for the imminent arrival of true artificial intelligence, things didn't pan out quite as expected. ALife did not see a breakthrough that produced artificial living systems, and for many observers it was a field that started suffering from a lack of focus, with too many intriguing areas to explore that were in the end perhaps only cute rather than meaningful. Perhaps not surprisingly, one of the barriers was the mystery of open-endedness, so seemingly effortless in nature, but so elusive in ALife.

Today there is a renewed sense that the questions underpinning much of the ALife field need answers more urgently than ever before, from the origins of life to what the phenomenon of life really means. Some of this is driven by advances in the ongoing search for life beyond Earth, in our own solar system and on the scores of planets discovered around other stars in the past few decades. These cosmic arenas are raising new questions about how we might even recognize alien life if we found it. Elsewhere, developments in genomics, chemistry, and biogeochemistry are energizing work on mysteries about the origins of life on Earth (or anywhere else). All of these lines of research bring our lack of good answers about life and its origins into stark relief, and back to the forefront of study. ALife researchers are wont to roll their eyes at the irony; if only there was a field that had tried this before.

That's another part of the reason why we're all here in Tokyo, to see if we can, to be blunt, find new life for ALife. It is difficult. In my summary on the second day, I show everyone the famous woodblock print by Hokusai of *The Great Wave off Kanagawa*. I tell the group that while we may all want to be secure on the slopes of Mt. Fuji—as it peeks above the foamy surf on the horizon—where we actually are is much more interest-

ing but dangerous. That's in the boat that is perilously close to being inundated by the geometric beauty of Hokusai's cresting wall of water.

There is a smattering of laughter. In many ways this is precisely where ALife has always been: paddling furiously and generating waves of ideas that either crash spectacularly back on you, or wend their way off to other fields of study, while the boat stays where it is. But as the days go by and I listen to the talks and discussions, one thing does become clear: There are features in the nature of information, and the connection between life and information, that might lead directly to an explanation for the dataome and its properties. There are also things about the dataome that might tell us about the nature of life.

One of the current efforts to understand what kind of phenomenon life really is involves understanding the networks that link organisms and information. In this context a network can represent a variety of things; a set of interacting chemical reactions or the interactions of organisms with each other, whether amoebae, insects, humans, or the peptide messaging of bacteriophages.

But how are networks different in living systems compared with those in nonliving systems? If these differences turn out to be systematic, this could be a way to discriminate between life and non-life and get at the secret sauce itself. This is something that Sara Walker is keenly interested in, and is what she and the physicist Paul Davies have called the "informational narrative of living systems."

## Emerging, Downward

Our recognition of that narrative has, in many ways, been around for a long time. Terms like "transcription" and "translation," "coding," "editing," "messaging," and "redundancy" are all happily utilized in any modern biology textbook, and wouldn't seem out of place in a handbook of computer science. But they're also all in service of the central dogma of molecular biology, which is that information moves from the nucleic acids in DNA to proteins, but not in the other direction.

There are in fact ways that information is able to exert at least some control in that other direction. I've already mentioned regulatory genes that produce proteins that affect how often other genes get to do their thing. And we know that epigenetic factors have influence on gene expression. So clearly something is up. A newer narrative, like that of Walker and Davies, tries to push us toward a theory of life, and is much more explicit about the nature of causal influences in the world. A causal influence is when, for instance, the properties of hydrogen and oxygen cause the structure and behavior of a molecule of water. That's a "bottom-up" causation driven by the fundamental atomic components of a system, and it's the way science usually thinks about the world: understand the underlying pieces and you can explain the big stuff—that's the beauty and power of reductionism.

But does that mean that "top-down," or downward, causation doesn't exist? In top-down causation, there would have to be something about a water molecule that causes property changes in the hydrogen and oxygen atoms themselves, which is not what we think happens for this simple structure. But move up to something like an icy snowflake and the signs of downward causation arguably begin to appear. Any snowflake has a strict sixfold spatial symmetry to its crystalline, regularized arrangement of water molecules. That symmetry can be attributed to the molecules' geometric properties (and the properties of oxygen and hydrogen). But in any given snowflake, the exact positioning of each molecule also depends on the positioning of the other molecules. It's the overall shape of the flake that effectively tells the molecules where they can or can't go.

An even simpler example is a wheel rolling down a hill. The properties of the wheel's individual molecules and the laws of gravity cannot easily explain the motion of those molecules. You can only explain those rules from the higher-level vantage point of the wheel itself, and in that sense the rules *emerged* from the formation of the wheel.

One of the best-known proponents of examining the world through the lens of downward causation is the South African physicist and cosmologist George Ellis. He argues that there are many systems that emerge in nature that cannot be reduced to understanding the interactions of their component parts. This is a pretty demanding statement. It doesn't feel like the way classical physics should work. Not surprisingly it's become known as the idea of strong emergence, and you have to believe strongly to go along with it.

Without going too far into the contentious discussions that swirl around this topic, we can still get a sense of what is meant. Ellis himself, in a 2016 book, states:

> The lower levels do the physical work, but the higher levels decide what work should be done.

An example is your computer or smartphone. Imagine what takes place when you press or touch the letter "A" on a keyboard or screen. A layer of software exists to sense this action, to pass it down to other layers of code, and to eventually flip the bits in microscopic transistors and memory cells. That process involves electrons physically moving through the metal and silicon substrate of a microprocessor. But the specific movement of those electrons is not something you could have predicted or even explained if all that you could see were those electrons and their individual properties. Even if you understood that a change was coming through the system because of pressing the letter "A," it might be extremely difficult to foresee the exact behavior of each and every electron.

The interesting thing is that life appears to do this too: in that example, what was going on with my finger press is part of a downward causation. You and I are made of cells, molecules, and atoms. These all have certain properties, but they all behave differently when they're part of you than if they were isolated. That is not to say that there's any change in their individual physics, but rather that the physics of these building blocks is not enough by itself to tell you all their properties and behaviors when they're part of you. I cannot predict that an atom of carbon in your foot will suddenly relocate to the other side of the room, but I can predict that you might walk to the other side of the room, taking that carbon atom with you.

In the last chapter I introduced the concept I called core algorithms, or corgs. What's interesting about corgs is that they too would be a very good example of strong downward causation, or what we might also call downward emergence—at least in the way I framed it. The flight corg causes atoms and molecules to come together into structures that, while heavier than the air they displace, are nonetheless capable of flying around. Could I possibly deduce this from knowing just the individual properties of those atoms and molecules? It seems unlikely without somehow simulating or tracking the behavior of every single atom with a fidelity that is, in effect, the same as that of the physical world itself. You'd have to build the world to explain the world, which feels a little absurd.

Whether or not strong emergence is the right story, scientists like

Walker and Davies, among others, are keenly interested in seeing how it might help us formulate a theory of life and aliveness. Specifically: top-down causation fits with the observation that living systems seem to be able to gain control over the very same matter out of which they are formed. This is a remarkable thing, as is the sheer complexity of the chemical networks embedded in life's causative behaviors (one reaction yielding new reactions, and so on).

That complexity is painfully tricky to decode into graspable rules and principles. For one thing, top-down causation might happen in nature via different mechanisms. Those mechanisms could depend on the physical sizes of things and the number of pieces involved. They are also likely to change according to context—walking an organism could be different from making an organism eat food. Perhaps, too, life distinguishes itself from other forms of emergent control in the way that its downward causation is distributed: the topology of the lines of communication from an organism's processes all the way down to molecules. That might be a hierarchical structure that looks like a tree's branches or roots, or it could be clumpy—a mix of "coarse" boxes and "fine-grained" detailed linkages, like the warehouses we use to gather our goods and supplies compared with the fine-grained arrangements of our home cupboards and pantries.

A further question is whether life is really a hybrid of digital systems (like DNA) and analog systems (like the way that cells sense chemical variations or electrical fields). Among all of these systems, connections, and information flow, information will have different qualities. Some will have less functionality and some will have more. Altogether it is an intimidating prospect to try to corral all of this into a new theory.

One promising inroad involves the observation that when life is alive (as opposed to being a corpse with its feet in the air), it can seem as if matter changes its state. There is a physical transition in the world, a fundamental alteration in the organization and hierarchy of matter, or what physicists call a phase change. Phase changes are critically important things in nature. We live with some of them all the time. When water goes from being liquid to solid, or vice versa, it is undergoing a phase change. Unseen by our limited eyes, the molecules of water are changing their organization and arrangement. In the case of water going from liquid to solid there is an increase in the order of things, a new regularity that comes with crystallization, a change in entropy and a change in the energy of the water. This behavior goes far beyond everyday phenomena. Phase changes are a key piece in the lexicon of theoretical physics. They often involve what is termed symmetry breaking. When water freezes it goes from a state with many symmetries to a state with fewer symmetries, because its molecules can no longer roam quite as free.

The phase transition proposed from non-alive to alive has to be more intricate and only truly visible when you investigate the hierarchy of information flow: tracing how the master algorithm of life (for want of a better term) reaches down through the layers it is made from to cause changes at the lowest levels of atoms and energy. Like the example of a carbon atom in your foot, how matter rearranges and moves itself in life is a consequence of the flow of information: the directives from upstairs that change the symmetries of matter and its freedom of state. But more than that, the transition from non-alive to alive (and from non-life to life in the origins of all living systems) may be about changes in *information-processing capabilities.* This is something we're actually quite intimately familiar with, because we can make observations about the macroscopic world (that food looks good; perhaps don't cross the street right now; it looks like it's going to rain), process the implications, and then direct our body's cells to behave a certain way.

Putting all of this another way: it's one thing for information to be encoded in matter, like the ink on a page or electrons in a silicon chip. It's a whole other thing if that same information can now exert influence over the matter encoding it. Printed letters on paper don't usually cause that paper to get up and walk around, but in living systems that seems to be precisely the kind of thing that happens.

One of the biggest challenges with the idea of information flow and control as a defining property of life (and a window into what could be new physics) is that it still tends to encompass entities or systems that we traditionally think of as inanimate. An example would be anything that a biological entity fabricates, because in that fabrication there is also a flow of information. It's just that it has a different temporal nature. Imagine that you're a carpenter and you decide to build a chair. Whether the chair is based on a set of printed plans or comes from your imagination and know-how, it only comes into existence because of the flow of information and an algorithmic "takeover" of matter. Of course, the chair doesn't seem like it's alive in the way that your arms or legs are, or a hopping bird is. But is that because of the particular clock that we're measuring things against?

Jim Crutchfield has also looked into a very closely related challenge: How can we actually detect and codify the emergence of any kind of complexity in nature? He's pointed out this kind of observation is inherently subjective. After all, I might think that a modern gasoline engine is enormously complex and confusing, full of valves, sensors, and computers. But a skilled mechanical engineer will see that engine as an obvious, relatively simple set of systems and interfaces. A way around this subjectivity would be to explicitly include it in the measurement, by evaluating the computational resources necessary in order to successfully understand a complex system.

That's quite the mind-bender. To phrase it another way: the process of measuring, mapping, and modeling a phenomenon is itself a phenomenon. Its success depends on the computational resources being applied: how much data is acquired, how much storage you have for that data, how much time you have for making inferences and estimations. Suppose I can only remember the last 10 numbers from an experiment or measurement. If the complex phenomenon I'm studying is manifested by the behavior of a thousand variables, it's going to be impossible for me to even *see* that complexity.

## Demons

Sitting in that room in Tokyo, in the winter of 2019, felt like being at a party that couldn't quite figure out if it was a success or not. Our excitement at being there mixed with friendly bewilderment about where we were headed. Despite that, there was a sense that some kind of convergence was afoot, a possibility of connecting the dots across an array of tantalizing insights. Some of those insights are coming from statistical mechanics and the linkage between entropy, energy, and information; and from the ideas of information thermodynamics, flow, and causation. So many gloriously teasing phenomena, yet that coveted convergence is proving awfully hard to grab onto. To use another analogy, it's like peering through the window of an absurdly well-stocked candy store, but you don't have quite the right coins to get what's inside.

There are areas with slightly clearer signs of progress, including work stemming from a surprising place: a peculiar but enormously powerful thought experiment published in the early 1870s by none other than the grand doyen of nineteenth-century physics, James Clerk Maxwell.

Maxwell had wondered whether the granular nature of matter, embodied triumphantly in statistical mechanics, actually left room for violations of the sacred laws of thermodynamics. He was particularly interested in the second law—the inevitable growth of entropy, and the need for work to be expended to create any kind of imbalance in an otherwise isolated, equilibrated system.

To test this Maxwell imagined a being (what came to be called a "demon" to connote mediation, not evil intent) with superb senses. That being could watch over a box full of molecules in a gas and see exactly where they were and how they moved. If the being controlled a door in the middle of the box it could, by snapping the door open and shut at the right moments, sort the molecules in the box by letting all the fast-moving ones accumulate on one side and the slow-moving ones on the other.

Since faster moving molecules give a gas a hotter temperature, the being could, without adding or subtracting anything from the gas (not doing any *work* directly on it), create a temperature difference between the two ends of the box. Not only could the temperature difference be exploited to drive an engine or process, the separation of molecules actually lowers the total entropy.

Lo and behold, Maxwell's Demon could seemingly violate the second law! Not surprisingly, physicists rolled up their sleeves and tried to understand what on earth this all meant. One possible explanation, as we've discussed, is that in any system described by statistics there's always a finite (even if tiny) probability of a fluctuation occurring that momentarily disobeys thermodynamic law. Maxwell himself thought of this: if, for instance, there were only a handful of molecules in the box it would stand to reason that sometimes the fast ones might be on one side and the slow on the other, just by chance. But the demon was an itchy sore in an otherwise happy bit of physics. Eventually, in 1929, that itch got some proper attention.

Leo Szilard was a Hungarian-born physicist who would play a pivotal role in the development of nuclear physics and the Manhattan Project (and the opposition to nuclear proliferation), as well as twentieth-century biology. But before all of that, he wrote his doctoral thesis on Maxwell's Demon and followed up with two critical papers. One of these was "On the reduction of entropy in a thermodynamic system by the intervention of intelligent beings." Published in 1929, it arguably made the first connection between entropy and information, predating Claude Shannon's work by nearly twenty years.

Szilard's insight can be paraphrased as follows: The demon can only segregate the molecules in its box by utilizing information. It has to somehow measure the molecules, to tell which are fast and which are slow. That information has to be instantiated in the world—whether in a demonic notebook or the neurons of a demonic brain. Even if the demon does no work on the molecules, it *does* do work to store and erase information, balancing out the decreased entropy in the box with increased entropy in information.

That also means that it should be possible, in principle, to convert information into useful energy. Remarkably, in 2010 a group of scientists in Japan published the results of a delicate and ingenious experiment in which they managed to demonstrate precisely this. Their actual experiment used a microscopic rotating polystyrene bead pinned to a glass sheet, buffeted by its surroundings and some carefully tuned electrical fields. But it's much easier to imagine the following: there's a ball bouncing on a set of stairs, but it's light and easily pushed this way and that by the air swirling around it. Left on its own it'll bounce up the stairs as much as down the stairs.

However, if you watched the ball and every time it went up a step you dropped a blocking wall onto the step below, eventually it should climb the stairs. Not because you're putting energy into it directly, but because you're being a type of Maxwell Demon: you're using information to allow the ball to only go in one direction. That means that the ball gains free energy, by climbing up the stairs. Incredibly, in the 2010 experiment the researchers could actually measure the equivalent of this staircase-climbing energy gain—evidence that information can indeed be converted to energy.

## Limits of Information

None of this means that you're getting something for nothing. The experiment I just described is a type of Brownian ratchet, a system that—on the face of it—seems capable of converting random, useless energy into meaningful work. Except in this case the catch is once again in the involvement of the experimenters themselves. They have to observe and measure. They have to instantiate that information and then move their equivalent of the blocking wall on the staircase steps. In other words, they have to do work, and in the end the second law keeps on holding steady.

Ideas and experiments like these have also led to answers to a deceptively simple-sounding question: are there absolute boundaries or limitations to information and energy? As I've described earlier, the basic efficiency of computation and data flow is an important factor in measuring the energy burden of our own information and our data. Although we are still managing to improve, squeezing ever more computations out of smaller amounts of energy, we're not yet at the efficiency level of a biological brain.

But setting that particular bar is actually quite arbitrary. I've also examined how the human brain, like most components of life, operates as well as it needs to but no better. This raises a question as to whether there is some ultimate limit to computation, a universal efficiency barrier set by the fabric of the cosmos itself, and by extension a limit to the informational nature of biology as well as machines. It turns out that there may be.

In 1961 the German-American physicist Rolf Landauer, then working at IBM, published a paper titled "Irreversibility and Heat Generation in the Computing Process." Much as Claude Shannon had done with information theory in the 1940s, Landauer had found a surprisingly deep connection between computation (meaning the manipulation of information) and thermodynamics—a connection that came naturally from ideas like Maxwell's Demon and Szilard's intelligent beings.

In essence Landauer's insight boils down to the fact that if you want to store or erase information, to change a bit in a computer (for instance) from a 0 to 1 or from 1 to a 0, you have to do work. For example, you may have to pass an electrical current through a piece of doped silicon to force electrical charges to move or change. That current is a flow of energy, some of which is transferred into physical things being shifted around. That process, according to our favorite second Law of Thermodynamics, inevitably generates a teeny bit of waste heat and can never be 100 percent efficient.

In slightly more sophisticated language: what Landauer realized is that if an *irreversible* manipulation of information is carried out, like the erasure of a bit in a system, that's just like lowering entropy in one spot. But the second law tells us that statistically, on the larger scale of a whole system, entropy cannot decrease. Therefore, when that erasure of a bit happens, something else needs to take up the entropy burden—whether it's the rest of the data processing machinery or the wider universe.

From this premise Landauer derived a formula for calculating the precise minimum amount of energy that is needed to erase one bit of information given the environmental temperature. That temperature matters because the environment is where waste heat has to go. To put this another way, this is the absolute limit to classical computation, the best efficiency you can ever hope to achieve in the universe, period. This *Landauer limit* is simply an energy, given in equation form as $kT \ln 2$. The quantity $k$ is a constant of nature called Boltzmann's constant. The $T$ is the environmental temperature in Kelvin, and ln2 is the natural logarithm of two.

Armed with a table of natural constants and a calculator we can quickly see the implications of this limit. In a room at 20 degrees Celsius the Landauer limit corresponds to an energy of about 3 zeptojoules, or $10^{-21}$ joules. If you had a magical computing machine operating at the Landauer limit of efficiency, it could flip bits at a rate of a trillion a second and consume only about 3 billionths of a watt of power (even less at lower temperatures). That's millions of times more efficient than today's conventional computers. You may also recall that in the humble blowfly visual system the energy cost of one bit of information has been gauged to be around $10^{-14}$ joules. While efficient, that's also some 10 million times more than the Landauer limit.

And before you wonder whether or not this limit also holds for quantum computers, as well as classical systems, the answer is that it probably does. Experimental work in 2018 seems to verify that even the quantum version of bits—qubits—are governed by Landauer's principle. The universe may just not let anything be more efficient than this. There is, therefore, truly a limit to the formation, flow, and erasure of all information.

## Darwin and Demons

How does all of this relate to the nature of life and information, and our conjectures about that? In part we're starting to see the fundamentals of how information *acts on the world* rather than how the world generates information. Maxwell's single-minded demon has more to tell us too. The ever-eloquent David Krakauer has described the idea of a "Darwinian Demon," or a "selective demon." Unlike Maxwell's thermodynamic version, a selective demon is (paraphrasing Krakauer) a demon that has enough intelligence to detect, memorize, and act upon biological variations in one generation of a species, in order to induce an adaptive distribution of genotypes and phenotypes in the next generation. We could say that the selective demon examines members of a species and decides their fitness, paralleling the way Maxwell's Demon examines the speed of molecules. More-successful organisms on this side of the box, less successful on the other side.

The point of this is not to suggest that selective demons exist (hardly!). Rather, this might be a useful conceptual tool because we can then ask questions about things like the mutual dynamics of such demons and the organisms they oversee. For example, we can ask what the comparison is between the rate at which the demon can gather environmental information, and the amount of information that organisms actually need to encode in their genomes to succeed. Both genomes and brains (and potentially dataomes) can be thought of as devices for gathering information that enables predictions about the future based on the recurring patterns seen in the past.

Casting our thoughts back to Jim Crutchfield's observation that ultimately you have to be more complex than the thing you're attempting to understand: a demonic selection principle also indicates that the complexity of organisms or species cannot be greater than the complexity of the environment (the demon) selecting them. But we know that organisms modify their environments, reshaping the selection pressures they feel. Perhaps, in some cases, this kicks off a virtuous circle. Just as a version of Maxwell's Demon can make a ball climb a staircase without directly poking it, maybe a selective demon can make an ecosystem get more and more intricate, climbing the staircase of complexity.

Other scientists have tackled related issues. One problem is that the more complex life is the costlier it is to maintain (an issue we'll talk about more in the context of Earth's specific history of life). Why then isn't all of life just single-celled microbes? An answer might be that more complex life can do so much better at decision making, at parsing environmental information to its advantage (overseen by the selective demons), that it outweighs the burden. Equally, and thinking back to our discussion of the compressibility of data, if complex life is better at the computations needed for lossy compression—throwing away irrelevant information—it can enhance its fitness without as much burden from carrying around all of that environmental information.

These proposals show a tantalizing kinship with our ongoing technological revolutions in machine learning and computation. We are intentionally building algorithms to take on cognitive tasks and massively extend our capacity to gather information about our environment, whether that environment is the natural world or the jungle of human data and human interactions.

All of the above parallels what I've described earlier in this book, on the nature of the dataome and its burdens and benefits. What's emerging out of ALife and our efforts to build a fundamental theory of life and intelligence is strikingly similar. It feels like the time is approaching for all of these pieces to slot together, for the science to undergo its own phase change.

## How We Think

But if complex systems require comparably complex measurements and minds to understand them, it's just possible that we're on a fool's errand. Evolution need not have selected our brains for their success at working out what life really is. We might not be up to the task.

There's still some cause for optimism. We are surely built according to the same principles that we're studying, so there's a chance that we've automatically kept pace with the problem. And we humans have also grown a dataome. With a dataome a species can flow information not just in and out of its own structures, but into a persistent and (in principle) shared entity—part of what I've described as a kind of holobiont. Maybe if we knew more about precisely how brains do their informational dancing, and get their computational heft, we'd also learn how the dataome amplifies us, and vice versa.

In neuroscience, figuring out exactly how our brains store and utilize information is still very much a frontier of inquiry. A key piece of that frontier is to do with what's called neural plasticity—the ability of neurons to undergo changes, again and again. One of the most remarkable stories in recent research on plasticity comes from the study of a piece of genetic code called the *Arc* gene. What makes the story particularly relevant here is that, unexpectedly, it brings us back to the behavior and capabilities of viruses.

The existence of the *Arc* gene was discovered by two laboratories led by Dietmar Kuhl and Paul Worley in 1995. In essence, when there is synaptic activity (at the chemical interface between two neurons), *Arc* messenger RNA is produced in the neuron and, to use the biological parlance, trafficked to the sites of activity, accumulating in the dendrites of the neuron cell and translated into other molecular structures like proteins.

A variety of experiments have shown that if you suppress the expression of the *Arc* gene you also impair the set of processes known as long-term potentiation (LTP). And LTP is all about how neurons change under stimulation, how they actually encode learning and information by growing new dendrites and strengthening or weakening synaptic connections to other neurons. Messing with *Arc* also messes with things like the formation of long-term memories.

Even though Arc is now often referred to as the "master regulator" of plasticity in our synapses, the precise mechanics of how all of this works are still unclear. But regardless of the inner workings, very recent investigations of *Arc* have started to paint a wholly unexpected picture for the origins and nature of this feature of brains. If you scrutinize the actual code of the *Arc* gene—about a thousand nucleotide "letters" in length—it becomes apparent that large stretches of this DNA look an awful lot like known viral DNA. Not just any viral code either, but code associated with ancient viruses and with how viruses encapsulate their DNA when infecting cells.

Being similar to viral DNA is by itself not a huge flag of interest. Enormous amounts of our genetic code, and the code of any animal, actually seems to have originated from viral additions throughout our deep evolutionary lineage. It's been estimated that as much as half of our genetic code is actually accumulated from the shenanigans of viruses over the past hundreds of millions of years. But the thing about the *Arc* gene is that it's present in everything from insects to humans, and the specifics of its code are similar to a modern virus gene called *gag,* a gene involved in building the particles needed for genetic transport.

When the *Arc* gene is expressed, it ends up making a protein that shapes itself into a little capsid inside the neuron. That, as its name suggests, is like a capsule, in this case like a virus-made capsule. Inside its parent neuron this capsid gathers up *Arc* messenger RNA that's floating around, and is then secreted from the activated neuron and invades other nearby neurons. The capsid escapes the originating neuron by becoming wrapped in a piece of cell membrane and pushing out like a bud that breaks free as a bubble, an extracellular vesicle. When the vesicle reaches another neuron, it fuses with that cell's membrane, releasing the Arc protein and RNA into this new cell.

Now, we don't know yet what the specifics are of all of the RNA that the Arc capsid gathers and then infects (for want of a better word) other neurons with. But by any standards this is an explicit transfer of information, of the genetic material expressed in one activated, stimulated neuron to others, strangely similar to horizontal gene transfer. Perhaps it's simply a way to prime other cells for making new dendritic growths, for LTP. But it could also be more complex, like students exchanging class notes on scraps of paper. Equally, it might be something else altogether, possibly even an adversarial move, sending fake news to the neighbors so they don't outdo you. Such a prospect is wholly unnerving: the idea that our 86 billion neurons might actually be competing with each other.

This may also just be the tip of the iceberg, both for how neurons interact, and how brains form and dissolve structures and encode information. It also points to the possibility of a communication network extending beyond the brain. Indeed, if you scrutinize the human genome you'll find about 100 other *gag*-like genes. These could potentially encode other proteins that form capsids that then go on to exit their parent cells in vesicles, those little budding extrusions of cell membrane that hark back to the phenomenon of exosomes we discussed earlier. What these could carry between the cells of animal bodies is impossible to know at the moment. But it could be part of a previously unseen communication network, a flow of information that has an uncanny resemblance to the mechanisms deployed by viruses, right down to stealthily hiding inside cell membrane vesicles and, presumably, evading the immune system in some way.

It's hard to resist making some further-reaching extrapolations, although these have to be taken with extreme caution. (If we've learned anything about terrestrial biology it is that things are seldom straightforward.) If we consider the tales of the great game of molecular permutations and bacteriophage peptide communication together with the story of the *Arc* gene, it's tempting to see a lineage of mechanisms all engaged in the flow of information, from atoms and molecules to cells.

From this flow come the physical changes in neurons as senses input information to an organism and that information is sifted and sorted. From that comes a collective, emergent intelligence—whether fly, mouse, or human. And from that comes computation, and prediction of the future, together with the generation of external, shed data, no longer encoded in heritable DNA (other than in certain epigenetic forms), but written in atoms and molecules. A modification of the environment and an information-driven climbing-of-the-stairs of complexity and evolutionary fitness.

In this framework, a species that *fully* externalizes its information, by making a dataome, is maximally extending its informational hierarchy. It is not only changing the state, or phase, of its own matter, but also more of the matter of the world around it. And, just like these hints that our own brains co-opt or evolve out of the tricks and tools of viruses, a dataome hacks the tricks and tools of biology. There are "infectious" memes; there is "reproduction" of data. A book or a newspaper looks a lot like a capsid wrapped inside a vesicle. We may like to think the dataome is all new, but it may actually be all the old tricks repackaged under the watchful eyes of Darwinian demons.

## After ALife

The flight from Tokyo to New York can take as long as 14 hours. Very often it follows a great circle around our spherical planet: up and across the Kamchatka Peninsula, and then the Aleutian Islands with their 44 active or recently active volcanoes strewn across an astonishing 1,200-mile arc, marking the imaginary division between the Bering Sea and the Northern Pacific Ocean.

I try to sleep, but my sleep-deprived mind veers from the mysteries of artificial life to the fact that I'm helpless in a darkened metal tube, careening across vast tracts of cold, empty ocean peppered with those Aleutian volcanoes. One conversation from the ALife conference keeps replaying over and over in my head. In the words of Sara Walker: What if we need new physics to understand life? If there is such a thing as a fundamental theory of the phenomenon we call life, it could be that it steps all the way outside of what we presently see as the canon of physics. It may well be that if there is new physics for life, it doesn't look like what we've come to think of as physics at all. After all, the transition from Newton's deterministic mechanics to an intrinsically probabilistic description of nature, in statistical mechanics and quantum mechanics, would've likely seemed very bizarre to Newton himself.

We've surely got some of the pieces, in information theory and thermodynamics, in emergence and complexity, and in Darwinian evolution and what rules we've decoded of molecular biology. There are also ideas for new ways of formulating known physics that seem better attuned to the requirements for explaining life. The physicist David Deutsch, in 2013, proposed *constructor theory*: rewriting our usual laws of physics in terms of *tasks* that are either possible or impossible—so something like the laws of motion get recast as what objects can and can't do. These tasks are really transformations or changes in the world, and generalize what we understand of thermodynamics, as well as information, taking us all the way back to the quantum "bits" that may be at the heart of reality. Constructor theory might just reveal the underlying impetus of self-reproduction and natural selection.

But to see all of these parts converge, to stick this all together, may be a struggle because we're still missing something—an insight that will be like the corner and edge pieces of a tough jigsaw puzzle.

I can't help but wonder if the dataome provides one of those pieces. After all, if some branches of life grow in complexity because it allows ever-better analysis of the environment, which provides ever-better future projection, surely the dataome is a revolutionary step. By encoding our information and our computations in structures external to our biological forms we've got a way to outstrip the usual constraints. But there are some nagging issues. The cost of the dataome is considerable; what it "wants" seems to pull against some of our interests. Is the song of the human-dataome holobiont really good enough to sway the selective demons?

As my eyes finally droop shut I'm thinking about the absurdity of flight. Of my fragile wet body inside an algorithmically directed machine, itself a consequence of the human dataome and four billion years of informational evolution. If there is a master demon of selection on Earth, it sure has a wicked sense of humor.

# 7
# LIFE MADE MACHINE

*Q: Are you alive? A: Perhaps in your fantasies I am alive.*

—interaction with the *Eliza* program created
at MIT by Joseph Weizenbaum in 1966

Stepping into Hod Lipson's laboratory feels like you'd imagine it to be going through the back of C. S. Lewis's deceptively ordinary wardrobe, entering a fantastical place that exists out of normal view, yet is also right beside where you're standing. For me, the journey to get here started innocuously enough: a walk from my office through a series of bridges connecting buildings on the campus of Columbia University on a wet winter's day in New York. All of which has only amplified the cognitive disturbance of where I've ended up.

Hod is a professor in the department of mechanical engineering, where he runs his Creative Machines Laboratory. It is a roboticist's dream world. The space is madly jammed with stuff, but with what you might call an emergent order to it. Shelving around the walls holds the remains of past experiments: There are insect-like walkers, some in multicolored hues. There are plastic and resin parts: robotic arms and claws, and things that defy quick explanation. There are small stacks of rubbery building blocks that are modules in a species of modular robots, and other components that look like lime-green Catherine wheels. There's also a row of eight-foot-tall green cornstalks. I don't get around to asking what they're there for. Perhaps it's simply to remind us of the organic world, even though they too appear to be artificial.

Then there are assorted 3D printers, and the frames of factory-like robotic benches for assembling projects. It's not unlike walking into a gallery in a natural history museum. A gallery for the fossil remains and working reconstructions from a particularly flamboyant period in Earth's evolutionary history.

I've arranged to meet with Hod in order to quiz him about the absolute frontiers of robotics and artificial intelligence. But I start by asking him about comments I've heard him make before, on the nature of biological life versus the world of machines. He smiles and slips into his easygoing Israeli-accented cadence. For him there's really no strong distinction. The substrates—the building pieces—may differ between organic life and other entities, but it's hard to see any real difference of *potential*—in the capacity for complexity, for evolution, and yes, for being that thing that we consider to be life.

Hod isn't alone. For example, a book written back in 1994 by Kevin Kelly, a writer, co-founder of *Wired* magazine, and futurist, was called *Out of Control: The New Biology of Machines, Social Systems and the Economic World*. It goes to town on the notion of a kind of unity among all complex systems. Biology turns into neo-biology (to use Kelly's term): there is a constantly unfurling engine of novelty, of the coevolution of matter, of pervasive technology. It's an industrial ecology, where the logic of biology is at the heart of the "next epoch" of machine-and-life-melded existence here on Earth.

The book is an explosion of ideas and wide-eyed eloquence, best read sitting down, and perhaps taken with a few grains of salt. Yet research is indeed underway on that industrial ecology. In 2019 the computational social scientist Iyad Rahwan at the Max Planck Institute for Human

Development in Berlin and his colleagues laid out an extensive argument for developing the study of machine behavior. They proposed treating today's machines and algorithms as if they are organisms roaming a new ecosystem.

None of these projections mean that we're necessarily quite there yet when it comes to our deliberate attempts at machine advancement. We haven't figured out self-replication in machines, which Hod feels is essential for a number of reasons: first, humans simply won't be able to babysit all of our devices. Right now, we're the maintenance crew on call, the fixers, tweakers, and decision makers. But what happens when there are more machines than human attention can possibly handle? It's predicted that in the 2020s we'll easily cross a threshold of 20 billion devices in the Internet of Things (the IoT). Autonomous, connected machines are also proliferating. The RAND Corporation thinks that by 2050 all road vehicles in the United States can, and perhaps should, be fully autonomous.

Self-replication would mean self-maintenance, self-repair, and self-improvement. And that leads to the second reason for it: without self-replication Hod believes that machines can't reach their full potential. That involves following the imperatives of survival and evolvability, achieving replication with variation—hooking up to the engine of Darwinian selection. The terminology is subtly different, though. Self-improving machines suggests something not quite like the natural selection that came before it.

Some of the work in the Creative Machines Lab is focused in this direction, even if it's for now manifested only in software. Hod and his researchers have a self-replicating AI (a quine, in language we used earlier), a machine-learning system that can learn how to pass on its own internal "weightings" to budded-off children AIs. These weightings are the numerical values describing its artificial neural nets, the somewhat mysterious numbers that define its operation. Because these numbers end up being correlated with each other, albeit in complicated ways, the whole net is quite compressible and therefore heritable. It's eerily reminiscent of biological DNA, which is arguably a highly compressed representation of an actual organism: the full complexity and colors of biological life really only emerge in the myriad ways DNA expresses itself in and interacts with the world.

But a goal for robotics is to move toward an actual physical ecology of synthetic life, even if mediated by humans for now. I say to Hod that my feeling is that, as with the puzzle of open-endedness and the growth of complexity in an imaginary demonic selection process for life, part of the answer must lie in the complexity of the external environment. We seem to agree on this point; there is a trade-off between our being in control and the complexity that our children systems can gain. Letting entities loose on the world is ultimately the route to exponential growth in complexity.

To begin to attain that growth, Hod's lab is also building modular robots that can self-configure and self-reconfigure. Early prototypes are the cube-like modules sitting dormant on the lab's walls, each with abilities to move, attach, and communicate with each other. Other groups around the world have been making similar experiments for a number of years now, going back to pioneering work like that of Toshio Fukuda in the 1980s with his "cellular robot." There are even mathematical frameworks that describe how to optimize the shape-shifting and locomotion of these LEGO-block species.

But there also has to be sophisticated artificial cognition. Robots with deep-learning software systems built in can form their own internal representation of themselves, starting from scratch. Hod calls this letting a robot "babble," like an infant human crawling around and bumping into things as it makes sense of the world. The lab has made robots that do this with no starting conditions and no initial information about how many arms, legs, motors, bits or pieces they have. One such robot, a black arm-like structure parked on one of the paleontological shelves, figured itself out in a matter of hours. It learned how to move, and in what ways it could move, in effect deducing Newton's laws of action, reaction, and motion all by itself.

Hod hands me a clear plastic container of what look like dozens of small pink cubes, each half the size of my pinky nail. These are next-generation modularity, a part of a project to create a robotic system that reproduces itself using materials and functional electronic, electromotive pieces all in the same mini-cube, or voxel, format. The parent, the common ancestor, is a human-made machine that performs a kind of 3D printing assembly of these pieces, using them to remake itself. And that child will then be able to remake itself, and so on.

The first goal is a self-replicating robot made of a million modules or building blocks. Such a machine is just the start for a world of what they're calling "digital matter": an environment in which machines, structures, even human food is built this way, not the analog way that we've always built everything that came before.

It really is a vision of a future of compounding exponentials, to use Hod's phrasing. There's something exhilarating and a little daunting about it all, if not anxiety-producing. You can begin to see how all the technological pieces of a revolution fit together: from the unexpected leaps in deep learning of the past decade to the comparatively simple generalizations of manufacturing embodied in something like 3D printing. It might not be so long until it's digital matter all the way down. Later, at home that night as I drift off to sleep, my mind is filled with babbling, self-reproducing machines swaying together in a field of pixelated plastic cornstalks.

But as I'm leaving the lab we get to talking about the notion of free will. At what point do machines not only do their own thing, but *want* to do their own thing—or is that even possible? It is the question of the nature of consciousness, and for Hod these are baby steps toward what could be an experimentally determined answer, by asking how we build self-awareness into a machine. Paraphrasing his words: Evolution has granted us the ability to imagine ourselves into the future, and that is very different than a bacterium simply following patterns of behavior that are expeditious in response to stimuli. Humans literally see themselves in the future. Why shouldn't machines eventually be able to do the same?

These advances and conversations bring us face-to-face with a question that I've been dancing around up to this point. I've described the human dataome using terminology and viewpoints lifted from evolutionary biology: cost and benefit, health and sickness, genes and selfishness, holobionts, and natural selection. But I've largely avoided just saying outright that perhaps our dataome is *already* another living thing.

In part that's because without deeper evidence this sounds plain silly. Indeed, over the centuries we've written so many speculative fictions on the notion of living, thinking machines that the discussion has become painted with the brush of fantasy, from the golem of Jewish folklore to the "hosts" of *Westworld*. Yet, as we've seen in the previous chapter, we don't actually have a good way of deciding what's alive and what's not. We also perhaps don't know what it truly means for a collection of matter to think, especially when that collection of matter is not thinking like us.

This is a place where I feel I may deviate from the stance of researchers like Hod (although they may all privately harbor similar ideas). I suspect that right now we do live in a world with more than one fundamental form of living system. And the dataome, encapsulating both abstracted information and the instantiations of machines, seems like a prime suspect.

Calling all of that a living system is quite the leap. Before I do jump off that particular cliff edge, it's important to recall some of the recent history that got us here. That history speaks to our struggles to imagine ways to discriminate between very differently constructed entities that nonetheless seem to do similar things in the world. It also helps us understand some of the reasons why our machines are made the way they are.

## Two Brains

It's impossible to talk about that history and the present nature of the dataome without invoking John von Neumann and Alan Turing. These two made fundamental advances and insights on an enormously long list of topics. And they did it all over a relatively brief timespan, from the 1920s to the 1950s.

Those topics included: set theory, ergodic theory, operator theory, continuous geometry, measure theory, computability, universal computation, quantum mechanics and mutual information, quantum logic, cryptography, game theory, fluid mechanics, cellular automata, morphogenesis, weather simulation, mathematical biology, and artificial intelligence. Only some of these terms may mean anything at all to you, but they've all helped shape the world you live in today.

In particular, the ways that we conceptualize modern electronic data, algorithms, and machines owes an enormous debt to what went on in the minds of these two individuals. There was never anything obvious or pre-ordained about how to make machines to compute or perform tasks. In a very real sense these rule sets have emerged with no more governance than was ever at play in life's four-billion-year story.

In 1945 the Hungarian-born von Neumann consulted for and published a report on a then-very-new computer called EDVAC (for Electronic Discrete Variable Automatic Computer) built by the University of Pennsylvania and installed at the US Army's Ballistic Research Laboratories in Maryland. Despite its gloriously unexciting name, EDVAC represented something of a revolution. Not because of its 6,000 vacuum tubes, 12,000 diodes, 56-kilowatt power demands, or memory units made from tubes of liquid mercury, but because of the way it was all put together.

Von Neumann's report described a fundamental reconceptualization of how computing should work: not only did EDVAC operate using binary arithmetic instead of what was then the convention of decimal arithmetic, but it deployed an utterly novel concept. It had *software.*

Instead of programming a computer by laboriously plugging and unplugging sets of wires and components, EDVAC embodied the principle of stored programs—now known as a von Neumann architecture. The program was held in the same electronic memory that held the data the program would crunch through, and that would hold the results of that crunching. It's a concept so familiar to us that today it's hard to imagine that this idea had to come from somewhere.

But for all of its innovative descriptions, von Neumann's report was just that, a report, not a meticulously thorough rulebook. That would come from elsewhere, because nearly simultaneously, British-born Alan Turing in London was working on closely related ideas that he published in 1946. These described, in rigorous detail, the full conceptual workings of a stored-program computer. In no small way, Turing's "Automatic Computing Engine Report" was the systems biology 101 of the electronic age.

Already in 1936, Turing had introduced a mathematical framework for what he called an "a-machine," or automated machine. We now call this a Turing machine. This imaginary device manipulates symbols on an infinite ribbon of paper divided into boxes or cells, with the ribbon chugging back and forth as needed according to a predetermined set of rules. In effect, the ribbon represented what we'd today consider a random-access memory, capable of being read from or written to anywhere along its length.

Turing was able to prove that this abstract machine could simulate the logic of *any* mathematically tractable algorithm. All algorithms—sets of rules for problem-solving—could be reduced to a sequence of machine instructions. He took this further to describe what we now call a universal Turing machine, capable of simulating any other Turing machine by also reading in the description of that machine. In modern terminology, this is what we would call a virtual machine. This conceptual move lives on deep inside how we construct our computer and robotic systems today.

There is no doubt that von Neumann was heavily influenced by Turing's incredibly advanced thinking, but both of them were instrumental in pushing the field of computation further. Von Neumann went on to consider seemingly outlandish ideas like how to build a reliable computing system from unreliable components. His inspiration? Biology itself, with all of its messiness and noise. He also helped invent the field of cellular automata, the discrete rule-following entities occupying a grid of cells that we came across before. Automata structures, behaviors, and rulesets form an immensely rich subject matter that continues to inspire ideas about the nature of life and computation. In his turn, Turing developed a mathematical and chemical model for how patterns might emerge in nature, from zebra stripes to the shape of bodies—the process of morphogenesis. In effect, he took very reliable equations and showed how their unstable interactions might produce the structures of the world. It was a sign that number-crunching computing could also generate elements of the complexity of nature.

Between the two of them they pretty much invented the fundamental principles—and philosophy—of computing as we have it now. Of course, it would be inaccurate to claim that they pulled all of these things out of thin air. The easy narratives we look for in history are seldom complete. Both Turing and von Neumann were informed and directed by the dataome of ideas on computing and mathematics and people before them and around them. These included Charles Babbage's mechanical computers of the 1800s and Ada Lovelace's mathematical leaps toward the very concept of algorithms and programs in the 1840s. Or Joan Clarke, who worked closely with Turing on codebreaking during World War II; and of course the work of people like Shannon and Tukey.

One of Turing's most original and intellectually provocative studies (albeit not as technically deep as his other research) was his 1950 work on what's become known as the Turing test. This is a test of machine and human intelligence, also sometimes referred to as his "imitation game." Turing begins his paper on the subject with the following sentence: "I propose to consider the question, 'Can machines think?'" Which would be pretty bold even today. In 1950 it was truly audacious.

In its original formulation the Turing test relies on judging the ability of unseen participants, whether machines or humans, to participate in natural language conversations. Turing's proposition was that this is a way to discriminate between thinking, feeling entities and artificial, automated devices. It's an idea that goes back a long way. Even in the 1600s the philosopher and scientist René Descartes mulled over the question of whether machines could be made in order to respond appropriately to human interaction via language. For Turing the question of whether or not machines could really think was better replaced by the question of whether or not machines can successfully imitate humans. He sidestepped the thorny problem of what it means to "think."

It's easy to level various criticisms at the Turing test. It seems plausible that an entirely mechanistic, unthinking algorithm could be designed to respond to any and all possible communications: fending off questions, making human-sounding mistakes, or using very natural turns of phrase to fool a human judge into thinking they're talking to another person. But the trick is in pulling off that imitation, because it's really not so easy at all. Much earlier, in the 1800s, Ada Lovelace (daughter of Lord Byron) had expressed how it might be unlikely for machines to accomplish this. Turing, in 1950, quotes "Lady Lovelace" as having written, "The Analytical Engine has no pretensions to *originate* anything. It can do *whatever we know how to order it* to perform" (her italics).

Turing admits that it's tough to provide convincing proof that thinking machines can ever be made, even though that's where he feels things will end up. He even proposes a direct forerunner of the evolution-inspired algorithms I brought up in the last chapter when talking about open-endedness, and an inspiration for Hod Lipson's "babbling" robots. Turing's idea is that a successful mind simulator is probably best made starting with a child's mind, or many such minds. These individual child-machine experiments would compete against each other to be educated, in order to discover what properties are fittest and should be inherited by subsequent machine minds.

The power of Turing's imitation game wasn't so much that it was a magical test that would set the bar for artificial intelligence, but that it framed the challenge in a much more useful way. What are the characteristics of human intelligence that are hard to imitate? What are the things that we do no better than machines? And, paraphrasing Turing's own words, what is it that prevents artificial minds, and perhaps less complex biological minds, from "going critical"—unleashing a cascade of thoughts and original ideas from the smallest of prompts?

Both von Neumann and Turing died when they were far too young, Von Neumann at age fifty-three (in 1957) and Alan Turing at age forty-one (in 1954). For von Neumann it was cancer; for Turing it appears to have been, tragically, suicide. Turing's story has become well known in recent years, and a key factor in his death was almost certainly the appallingly brutal legal and medical treatment he endured because of his sexuality. Humans can be a remarkably self-defeating species in their behavior toward one other.

Indeed, if both Turing and von Neumann had lived into the 1980s, I wonder what our current state of artificial intelligence research would have been. What would our perception of the promises and perils of machine intelligence have looked like? It is possible that we would have arrived at a version of today's machine learning much earlier and have felt its impact across our societies far sooner.

Today it's hard to miss the debates and opinions swirling on where our machine intelligences are headed and where they're taking us. Some of this discussion is spot-on: deep-learning systems, and techniques like adversarial learning, can be potent, disturbing tools for aiding or undermining human behavior. They might not really pass the Turing test, but their imitation games are good enough to confound us. Similarly, our increasing reliance on machine decision making that is largely opaque to us in terms of its rationale surely comes with potential dangers.

But there is also enormous hype and overstatement in these discussions. The fact that so many of us—scientists and otherwise—conflate artificial intelligence with machine learning in our language choices is particularly misleading. We definitely have machine learning, but it's not clear that we have anything approaching true artificial intelligence—as in Turing's "thinking machines." It's equally bad when some of us generate breathless conversations about utterly hypothetical concepts like "superintelligent machines"—as if these represent anything manifested in the world, or indeed anything that will ever be plausibly manifested, when the answer is that we simply don't know.

Instead, I think reality is subtler and vastly more interesting. The history of von Neumann and Turing speaks directly to this fact. The extraordinary richness of what they saw emerging in mathematical structures and logical puzzles was a signature of a phenomenon that we are already deeply entwined with, a phenomenon that may make any concerns or musings about AI fade into inconsequential noise: the possibility of an alternative living system already here on Earth.

Turing and von Neumann helped seed a conceptual framework of computation and intelligence that has served us extremely well. But it's also—like our language of organisms—predisposed us to seeing the world a certain way. Computers and software might be appealing metaphors for how life works; the code of the genes, the algorithms of evolution. Yet there is no incontrovertible proof that these are really such a good fit.

The converse is true too. Applying the patterns of Darwinian selection and evolution to our external world of data and machines, embodied in the dataome, may have some convenience, but it may not be quite right. While the age of robotics and digital matter is unfolding through work like Hod Lipson's and many others, we don't really know if the existing paradigms of terrestrial biology can, or should, describe all of this.

The simple truth is that there's no guarantee that what comes next will look like anything that came before, or that it won't force us to completely reshape all of our ideas about life, intelligence, and consciousness.

There's also no guarantee that a transition hasn't already long since happened, but we've been slow to realize it.

In that spirit, let's spend some time with a special thought experiment.

## Metalworld

From a high enough vantage point in space, Earth appears as a lonely globe painted with an exuberant spread of shifting blue-white textures. Raining onto this sphere is a barrage of solar photons—part of the fearsome and ceaseless stream of energy flowing out from the Sun and across the solar system. Any world intersecting this flow casts a radial shadow into the cosmos, a momentary tube of darkness reaching onward to infinity before the planet's orbit carries it to the next spot and the next after that. On a rotating world like Earth there is also a constantly moving edge between dark and light, night and day. This edge is called the terminator, and at this particular moment it is racing across the Atlantic Ocean at a thousand miles an hour, heading toward New York City.

The coming of the terminator triggers a flurry of activities in that planetary metropolis. In a network of tunnels and raised causeways a population of hollow, segmented metal tubes pick up their pace in a whirl of back-and-forth movement. These tubes race between sets of fixed points and concrete and steel structures—some aboveground, some underground. Each set of tubes pauses for a minute or two at these places before accelerating on to the next hub.

Electricity powers them, fed into metal rails that line their paths, and carried through parts of a vast and tangled network of metallic ropes. These ropes themselves connect into other networks and crisscrossing grids of cables and conductive metals, eventually terminating in widely placed clusters of large sheltering structures. Here spinning turbines and giant magnets convert the thermal energy of combusting hydrocarbons, among other sources, into flowing electrical charge.

In another, even more complex set of patterns, hundreds of thousands of small metal-and-glass pods balanced on dense rubber wheels are springing into motion. Many start out sitting on top of specialized plateaus of pulverized rock and stone, or inside very similar structures of stone and concrete. Others start along parts of an enormous network of smooth, ribbon-like mineral surfaces. Hardened mixtures of granular materials of varying dark gray tones stretch across the landscape, sometimes following its undulations, sometimes cutting straight through hills, rivers, or valleys.

These wheeled pods appear to be aware of each other and of numerous waypoints along the ribbon-like surfaces, indicated by abstract symbols planted on metallic posts or electrically powered light sources, blinking in red, amber and green. Swarming across dozens upon dozens of miles of the ribbon network, these objects alternately spread and coalesce around the city and its environs. They temporarily synchronize in regularly spaced chains, then split into varied configurations that disperse and reassemble again and again across the terrain.

In the air are streams of winged objects, some appearing from the deep night of the western continent, some from the bright side of the terminator—racing to the surface from high above the glistening morning ocean to the east. On the ground, tens of thousands of lights are turning off after a night spent illuminating patches of land or the interiors of the boxy structures of rock, steel, and wood that sprawl across the landscape. Most of these structures—some are modest, some are enormous towers reaching high into the sky—contain a variety of mechanical engines that hum and click as they heat or cool the air and water that circulates through them.

And, although impossible to see from above the planet, within an extraordinary gossamer-like web of copper veins, glassy fibers, and invisible beams of electromagnetic radiation, this metropolis is pulsing with the flow of data. Trillions upon trillions of staccato blips stream second by second through metal and air. They come so fast and so numerously that they merge into one great hiss of white noise, like every wave crashing simultaneously on every shoreline on every continent.

This one city is mirrored across Earth in other cities and agglomerations of structures, sometimes on an even larger scale, more often on smaller scales. As the Sun's terminator races around the planet similar patterns of activity are triggered, before tapering off in the depths of night, only to happen again with a new dawn.

Gazing down from a high orbit we see that all of these phenomena exhibit properties that are compellingly organismal in nature. The longer we watch the more this becomes apparent. There are hierarchies and relationships that might easily be attributed to a long history of Darwinian selection and evolution. Phylogenies—family trees—might be assigned to any group of what look to be species, from the metal tubes racing underground or in the air to the block-like, rooted objects that cluster according to geography and geophysics, hugging sunlit hillsides or nutrient-rich river banks and shorelines.

In the noisy flow of data we see signs of what (from our wet biological perspective) we might call gene expression. Information is being transformed from persistent, but not truly immutable, storage to instructions and new physical forms as a hierarchy of systems combine minerals, metals, and hydrocarbons to build entirely new, often duplicate, structures and devices. Yet these do not stay the same. With time there are changes and the spreading of shared characteristics. The arch-rule of Darwinian selection is that there be variation within a species and that variation be heritable. This certainly seems to be the case for Earth's metalworld.

There are also hints of what we might see if we waited even longer. There are clashes and crashes: function and dysfunction in the face of both the external terrestrial environment, from winters to summers and their storms and surprises, and the environment of all the other organism-like entities. There are also signs of parasitism. Machines and structures host other machines, and sometimes those attachments seem to subvert the function of their hosts to their own ends—gobbling up power or causing a physical, anatomical rearrangement to better accommodate themselves.

Finally, if we peer really closely we might discover that there is another realm integrated with the function of the metal. This one is filled with organic chemistry. Much of that chemistry is in the form of individually mobile sacks of water and proteins that appear to play a support role to everything else. The sacks are often temporarily embedded in spaces inside the machines and other pieces of a place like New York. In many instances they are engaged in the reproductive processes of the metalworld. They also seem to participate intensely in the flow of data—acting as conduits and connectors between the inorganic units of the world. They absorb, generate, and react to data relentlessly, and they are present everywhere in this ecosystem.

As time goes by, the organic sacks appear to take on a variety of additional roles for the inorganic world. Sometimes they are an immune system, fending off rot and rust and fixing problems. Rebooting devices and reconnecting power. Signaling to call together groups of specialized forms and spares to repair and rebuild. In other instances, these organic units participate in metabolic processes, seemingly making decisions on refueling and supplying engines and other machines. They also busy themselves seeking out and experimentally evolving new materials and energy sources in support of the metalworld.

These organic blobs are so prevalent, and so much of their behavior seems to have a degree of agency, that it's tempting to think of them as another "ome." A symbiotic, perhaps coevolved collective, part of a holobiont. And perhaps most striking of all are the examples where these organic entities are increasingly subservient to data and structure in the metalworld, whether pausing to be fed data or being co-opted into furthering the evolution of the metal, by building and nursing new forms. The hierarchy of information flow, top to bottom, metalworld to meatworld, is apparent.

If we could peer into the cacophony of data in the metalworld we would even find algorithms that have absorbed the most unique qualities of agency of the organic sacks. These net-like algorithmic processes assign near indecipherably complex and subtle values to their nodes and connections, and learn how to analyze and predict aspects of the present and future with uncanny accuracy.

Enough. At this point I've probably overstayed my welcome with the idea of retelling the story of modern Earth as a planet of machines and data. Numerous science fiction authors will be rolling their eyes at my lack of originality. But the point of playing this game for a couple of pages is important, because it shows how incredibly easy it is to reframe our world this way. It seems . . . almost natural.

There is of course a very long step to take from that appearance of naturalness to making a serious claim that the world we live in is fundamentally different than we think it is—that there has been an emergence and evolution of inorganic living systems on the planet. But are we just being parochial in our ways of looking at the world?

Consider the overarching conceit that if we magically made "us" (organic life on Earth) invisible, some alien visitor might still think that this is a living planet based on the activity of the "metal." On many levels it seems like they could easily draw that conclusion. Across all of the visible inorganic structures there is evidence of growth in habitats and niches, there is diversity of forms and features, and there is certainly evidence of variability within species (think of all the variants of bikes, cars, airplanes, ships, houses, roads, factories, smartphones, computers, washing machines, and bathtubs).

Watch for long enough and there is definitely something akin to evolution being driven by a variety of selection pressures. In the space of about three hundred years the planet has seen the emergence of new species with new metabolic processes, from steam engines to solar electric power. These strategies for energy use have, much like terrestrial biology, reshaped the chemical balance of the planet. The atmosphere contains much more carbon dioxide than it did before these innovations. There has been extensive reworking of geographical areas as mining and refining has taken place. And oceans and atmosphere have experienced the addition of chemicals and hydrocarbon particulate matter (such as microplastics) that are entirely the product of the metalworld metabolisms.

There is even evidence of what might be construed as gene flow—the transfer of genetic variation from one population to another. We see this kind of migration-driven change as certain kinds of machines or technology finally arrive in new regions of the planet and blend with the local fauna. Some examples are the flourishing of mobile phones across Africa and South Asia, or the migration of technological trends and designs across the planet from the suburban origin points of Silicon Valley.

The related emergence of increasingly complex microscopic computational systems has taken just a few decades. Each generation of computing power yields to more powerful systems constructed at ever finer scale, pushing against the tyranny of entropy that we encountered several chapters ago.

For much of their brief history computers have followed Moore's empirical law with a doubling of density in the number of logic gates (transistor switches) in semiconductor chips every year. This kind of exponential growth is a consequence of compounding enhancement. A few percent change adds to the previous few percent change and so on. It is not unlike a biological innovation that invades an entire ecosystem. With this astonishing growth we've also seen the rise of quite radically different instantiations of machines and data, from the relatively slow transition of punch card readers from textile looms to digital computers, to the extremely rapid emergence of deep-learning algorithms prompting new physical chip designs and computing strategies.

Our machines may not have yet attained full open-endedness by themselves, but with us along for the ride they certainly have. For an imaginary external observer, it might seem that the metalworld and the meatworld are neither very different from each other nor truly separable. Both are evolving, living systems.

And, in the language I introduced of core algorithms—information algorithms encompassing parochial implementations like DNA or genes—it's surely not very surprising that the phenomenon of life on Earth has found yet another way of implementing itself.

## Hidden Layers

Statements like these, even when made cautiously, rightly cause some pretty ruffled scientific feathers. We should always be concerned about overinterpreting similarities between phenomena when those similarities could just be superficial in nature. But how do we find out the truth?

Like what Hod Lipson and his laboratory collaborators and other roboticists and AI experts have done, one line of attack is to pursue direct experiments in machine evolution. Eventually we may satisfy ourselves that a machine species has emerged that is as much alive as we are. Alternatively, like the scientists examining machine behavior in the existing ecology of the metalworld, we can follow Darwin's example: studying both the large-scale patterns of the world and the patterns in the minutiae. The challenge, which can't be overstated, is that we're not even sure we understand what all the patterns should be or what they're telling us.

History has plenty of signposts that might offer insight, especially when it comes to phenomena that are nested inside of each other, obscuring the building blocks. The discovery of genes, and eventually their physical construction in DNA, are good examples of how difficult it can be to find the inner mechanisms of what's in front of us—or even deduce that there have to *be* inner mechanisms. And for ages we twisted ourselves into knots trying to divine the rules explaining how compounds behave with each other—what we'd today call chemistry. Transformations of substances, even hypothetical transmutations, made for some pretty hairy theories about the nature of matter. For a time people believed that all flammable compounds somehow contained a substance called phlogiston that could be released in fire, or that an element such as the metal lead was measurably more sinful in its drab greyness than gold with its virtuous, god-given luster.

The first semi-scientific identification of a pure element (other than a number of metals) wasn't until 1649, when the German merchant and wannabe alchemist Hennig Brand refined white, waxy phosphorus from 1,500 gallons of "beer-drinkers' urine," and recognized that it was a unique substance immune to further efforts at refinement. Possibly to the relief of his neighbors.

Then it wasn't until the 1770s that oxygen was finally understood to be a fundamental element. The concept of a periodic table of elements only came a full twenty-five years later, proposed by Antoine Lavoisier with critical help from his wife, Marie-Anne Paulze Lavoisier. That fairly crude table contained a total of 33 elements (compared with the 118 now discovered or synthesized). One challenge was the simple fact that substances composed of many unseen elements can be just as chemically reactive and as complicated as substances composed of pure elements. So, it was awfully hard to claim that there was something special about any particular isolate. We needed to let go of a number of preconceptions about the nature of matter in order to see the deeper encoding in indivisible atoms.

Similarly, our current inability to pinpoint all of the underlying principles or patterns of life, other than seeing hints in information flow and thermodynamics, may be hindering the way we see machines and the dataome. We need a checklist, as with the elements of the periodic table, to gauge how the metalworld really compares with biological systems. It's a big problem. I can't possibly claim to do it justice here, but we can take a look at some small pieces.

For example, if the metalworld really has its own natural selection and evolution, does it too have a universal way of encoding forms and functions across generations? Do robots have a version of DNA?

## Heritable Metal

A fascinating property of all machines, no matter how simple they are, is how they physically encode meaningful information. A bicycle is a partial repository of the information for making a bicycle. It may not explicitly encode the composition or manufacturing processes of its alloys and plastics, but it certainly encodes how they should be assembled. It also effectively encodes information about humans themselves, our anatomy, our sense of balance, and our ambulatory limitations.

If you examine that bicycle's lineage you'll be able to map its origins back through a hierarchy of other machines that have forged, lathed, molded, and welded its pieces (as well as, in this case, a likely contribution from human hands). Those other contributors (it's tempting, but premature, to call them the equivalent of enzymes and molecular chaperones) encode more of the information required to manifest a physical bicycle. And, of course, the entirety of the information necessary to make a bicycle will exist somewhere in the symbolic representations of the dataome, on paper or in electronic form.

But, critically, a bicycle cannot self-reproduce. That rusty specimen in your backyard will not one day produce lots of baby bicycles. That makes this kind of machine appear quite different from biological entities—all of which carry in their individual DNA the information necessary to reproduce themselves, as well for replicating the information in that DNA.

The idea of closing this loop for machines by incorporating fully embodied self-information hasn't escaped researchers' attention. Hod Lipson's steps toward self-reproducing robots are one such effort. Elsewhere, in 2019, a group led by Robert Grass at the Eidgenössische Technische Hochschule Zürich came up with a way to place DNA inside tiny nanoscale glass spheres that could then be mixed into materials like polyester, epoxy, and silicone. Several thousand strands of artificially constructed DNA can encode information about the very object that they're then incorporated into. For instance, a contact lens can hold the wearer's prescription details, or 3D-printed bunny rabbits can contain DNA-encoded instructions for their own ears and whiskers. In this case, though, the use of DNA is simply a convenience. Another group of typesetting molecular structures could work, too.

But that of course wouldn't make a machine like a bicycle capable of self-reproducing, because it doesn't have ways to replicate by itself. As I've discussed, though, the process of self-replication in biology is itself complicated and perhaps not entirely dissimilar. A multicellular organism may start with a single cell, but its full assembly involves an incubating system and a multitude of steps beyond basic cell division.

First, there is cellular specialization. Initially identical pluripotent stem cells can end up making quite different kinds of cells. There are also embryological "body plans," where the assembly of cells into sequences of intermediate structures involves the properties of transcription factor proteins like those expressed by the so-called Hox genes. There are certain compounds called morphogens, whose spatially varied secretion sets up chemical gradients that guide cellular assembly (these fascinated Alan Turing). For most larger animals there are complex juvenile stages where organisms continue to grow and learn to interpret their sensory inputs and their own mechanical properties. And, as I've discussed, a creature like a human can only function fully if complemented by its microbiome, a microbial army accumulated from the environment.

Something as complex as a multicellular creature doesn't just spring into existence from a magic seed. It is assembled out of morphing, interacting, interdependent pieces that all have environmental dependencies. We might blithely call that self-reproduction or "self-assembly" but that seems to be an oversimplification—it's self-assembly that relies on preexisting entities (parents, gestational sacs), passes through different stages in time, and is strongly influenced by the surrounding world.

In this sense the replication and assembly of machines, whether bicycles or factories, is actually perhaps less different from biology than it might at first appear. It requires many mechanisms, stages, and environments. To be clear, this is true for the *present* metalworld. I'm not explicitly including pure self-replicating, evolvable machines, which for now are still an experimental dream. Instead I am talking about the "natural" machine world, the one that has emerged all around us.

There are, of course, still many specific and intriguing differences between machines and biology. Compared with biology, there is a much greater separation in space and time between a machine's traits (or a phenotype) and its underlying information storage systems (a genotype). Our bicycle doesn't actually need to contain *any* of its own information (although it naturally does) *in situ* at the same time that you're riding around on it. Whereas for biology that spatial and temporal connection is vital to Darwinian evolution, so this difference is worth looking at more closely.

In biological life genes can be steadily expressed into proteins. In a mammalian cell the process of transcription of DNA into messenger RNA can take about ten minutes, followed by perhaps a minute for translation of the messenger RNA into a specific protein. These molecules are spooled out of complex ribosomal molecular machinery. Their production and folding is impacted by the interplay of DNA's own undulating, vibrating molecular structure with an environment that is itself undergoing continual tweaking as these gene-generated molecules come and go. All of which makes for a physically close-knit, tightly bonded community of players.

That bonding continues up the chain of complexity. Cells have survival dependencies on other cells, and on their mutual environment. That environment may contain physical and chemical hazards, but just as our skin protects our innards, cells can improve the conditions for other cells. Further along the chain, for the cellular groupings that are multicellular life, organisms have complicated survival dependencies that play out across multiple timescales, from hours to days to years.

Critically, the chain transmits direct survival feedbacks that enable the mechanisms of selection to reach all the way back to the genes themselves—to the individuals and lineages carrying them. Those feedbacks can quickly stem the flood of a severe dysfunction or gently coax variations that serve successful propagation into the future.

In other words, the core information, encoded in DNA, is completely engaged with cellular or organismal operation and success or failure. By comparison, if a machine experiences challenges or successes in the world the feedback to its core information may be quite indirect. If your non-internet-connected refrigerator breaks, you probably don't perform a detailed inspection and send a technical report to the manufacturer so that they can address some minor design flaw. You might not buy that brand again, but in the grand scheme of things that's a soft, long-term form of feedback. The refrigerator lineage does not end with you. Even if it is hooked up to the internet, there is no guarantee that the fridge's failure report will change anything in future designs. Similarly, if the wheel on your bicycle breaks, that may never influence the future design and manufacture of that species of bike.

I can always keep the master plan for something like a bicycle or robot securely locked away, no matter what these machines do or "learn" in the world. Even if there is feedback toward updating and improving a blueprint, the reset button is always an option. Consequently, we might be forced to say that natural selection as we understand it really can't apply to machines—at least not until we unleash those hypothetical self-reproducing, adaptive, variable robots. In other words: machines that copy everything that biology does.

But what if, to steal a clichéd line, life has found a way? Or, in the language I've introduced, what if the core algorithms, the corgs, have found another way to utilize elements of natural selection to spread their implementation in the world?

For many if not all biological species the relationship between traits or phenotypes and the genotypes are decidedly complex. There are also cases where some of the selection that acts on the fitness of phenotypical traits no longer directly impacts the survival of individual genotypes, but rather is mediated or transmitted through another system—namely, us.

Human agency can massively subvert the flow of biology. Think about dog and cat breeds, plants, and domesticated livestock. We're so intent on making these organisms adhere to aesthetic ideals or commercial standards that we force uniformity on their genotypes. We completely circumvent the usual "rules" of fitness and success. Fitness for human purposes can be very different from fitness for survival in the wild. Similarly, engineering uniformity and yield into crop species serves our immediate purposes, but produces vast monocultures that lack the variations that might protect them from disease or predation. These species persist, even flourish, because their fitness is measured by *us* and we then exert selection forces that can outweigh those in the rest of the environment—by deploying everything from pest control to vaccinations.

One of the most vivid examples is that, at this very instant, there are about 23 billion domestic chickens alive on Earth. That's roughly ten times more than any other avian species. There is also a perverse population dynamic for these creatures whose ancestors were jungle fowl, and whose heritage traces back to the avian dinosaur species that survived the Cretaceous-Paleogene mass extinction 66 million years ago. Every single year about 65 billion chickens also perish to feed humans or because of other attrition. That's a massively lopsided death rate compared with the steady-state population number for a species with a natural lifespan of around seven years.

Yet if an alien visitor were to attempt describing the world from the perspective of chickens, as I did for machines earlier in this chapter, they might easily conclude that chickens, not us, dominate the world. Seen through the cold eyes of evolutionary success and selfish genes, chickens are one of the planet's greatest triumphs. But we know the immense price they've paid in terms of individual suffering and stagnation.

There may not be so much difference at all between this phenomenon and the situation for machines. Genotypical success can be effectively governed by human agency. The chain between genes and selection forces can be routed through other species, and the genes themselves can, in effect be held separately, securely, seemingly controlled outside of typical Darwinian channels.

The answer to whether or not robots have DNA appears to be that they have something that accomplishes most, if not all, of the same function in the world. But it's differently implemented. Their core, heritable information does not need to be held in individuals, or even within a given species. That information is dispersible, although often localized in the dataome. Consequently, and not surprisingly, humans are an integral part of the lifecycle and evolution of machines as they presently exist.

## Machine Sex

In life's history on Earth the emergence of sexual reproduction over a billion years ago (sometimes labeled outcrossing to emphasize its difference from asexual reproduction, or self-replication) stands as one of the greatest innovations. We've already encountered the phenomenon of sexual selection in evolution, hinting at how complicated things can get. The story of why sexual reproduction has evolved on Earth is itself hugely complicated, and I think it's fair to say that there is still no simple consensus view. Part of the conundrum is that sexual reproduction results in slower population growth than asexual reproduction, and it's also inherently costly and competitive—each parent's best interests are, at least superficially, to propagate their own genes and not those of their mate. And often male individuals are incapable of bearing offspring themselves (useless things). Clearly though, somewhere down the line there has to be a terrific evolutionary advantage to outweigh these, and other, apparent handicaps.

One advantage comes from the incredible variational mix-and-match of sexual reproduction, the randomized fusion of sets of genes with potentially different encodings and an increased efficiency of gene repair and weeding out deleterious mutations. Another arguable advantage has to do with the intense competition between hosts and parasites. This is sometimes referred to as the Red Queen Hypothesis, where, as in Lewis Carroll's *Through the Looking Glass*, you have to run as fast as you can simply to stay still. On the face of it, the rapid genetic variation enabled by sexual reproduction in a population can help a species outpace (or at least keep pace with) parasitic organisms.

A stunning laboratory experiment reported in 2011 tracked the evolution of 70 generations of the tiny nematode *C. elegans* together with a pathogenic bacterium. The *C. elegans* were genetically manipulated to reproduce both sexually and asexually. In the end it was the sexually reproducing nematodes that won out, supporting the basic notion of the Red Queen Hypothesis for the evolution of sex.

What does this have to do with machine life? Well, for one thing, if we want to experiment with self-reproducing robots, as Hod Lipson and others are doing, we probably need to consider what those machines will eventually face "in the wild." There are certainly machine pathogens, in the form of software viruses but also in the form of chemicals and erosive processes, like rust and plastic-degrading pollutants. Whatever our machine-children look like, they may need the advantages of sexual reproduction. What form that outcrossing takes—two parents, three parents, a thousand parents—remains to be discovered. But we might also ask whether or not the present metalworld already involves some kind of sexual reproduction for machines. And this is where things get interesting.

Let's circle back to the good old bicycle. A bicycle's physical parents are different species altogether, some combination of welding robots, metal fabricators, assembly lines, and humans. New bicycles do not appear after a pair of parent bicycles go out on a hot date. Nor do they appear spontaneously, as a form of asexual reproduction. The closest we can get to parental seeds in the biological sense is if we look at the design documents and spreadsheets that, in most cases, live on a company's computer servers or on the desk of a designer. And, as for most machines and manufactured objects, the plans for a specific species are amalgams, hybrids, and blends of other variants. A wheel from here, a handlebar from there.

The same is basically true no matter what component of the present metalworld you look at. Automobiles start off with parent designs and plans that pull together previous ideas. Software and algorithms are more often than not the informational children of earlier software and algorithms, merged, mutated, weighed against each other. Although humans have played the central role of matchmaker-partner, that is often a pretty mechanical, uninventive process. You take what worked before and add it to other stuff that worked before.

In other words, there is outcrossing in machine reproduction, but it happens far up the spatiotemporal chain. A designer might cobble together the plans for a new species of toaster sitting in an office in Manhattan, but only see it manufactured in China three years later. At the same time, a machine species might be replicated millions of times, all from precisely the same outcrossing event. To summarize all of this: machines deploy both sexual reproduction (of a sort) and asexual reproduction (akin to cloning).

## Connecting the Dots

In the first few chapters I told a story of the human dataome, its history, and where it's at right now. A big element in that story is the impact of the dataome on us, and on the planet. To explore that impact I talked about what appears as function and dysfunction in the dataome. On the side of function we looked at the amazing advantages to be had from information that persists across time, as well as extraordinary discoveries showing that our symbolic representations of information, our written languages, can change the structure of our brains. But on the side of dysfunction we saw how exponential technological growth and resource use is pushing hard on the natural environment, contributing to climate change and all that follows. We also saw how tools forged, in part, from the landscape of information theory—like adversarial learning—may degrade, or sicken, parts of the dataome by reducing the amount and accessibility of meaningful information.

The possibility that we already live in a metalworld, on a planet with more than one fundamental form of living system, isn't just consistent with that earlier story, it could help explain it. A metalworld goes one step further even than the provocative notion of a human-dataome holobiont. It now incorporates the physical manifestations of the dataome: the machines, including all the ways they are instantiated, from tiny sensors and devices peppering the world to robots and computers, all hooked into continent-spanning systems of power and transport.

A simple and direct consequence of a metalworld is that we find ourselves experiencing all of those evolutionary experiments playing out around us and pulling us in, even if we don't at first make the intellectual connection. The exponential growth of unsustainable technology? A perfectly natural result of the interactive dynamics of living systems, oscillating around some as-yet-unseen equilibrium or balance point. Or possibly lurching off to an equilibrium that also sees our species going extinct. The data that seems diseased or dysfunctional for human purposes? A mutation, an evolutionary experiment, whose longevity is yet to be determined.

Here is a new perspective on our place in the world. It reaches far beyond any parochial notions of human economics or social behaviors. It is a way of looking at the planet like those imaginary visitors seeing the metalword from orbit. It is fully holistic, incorporating the biosphere and the technosphere in one integrated, pulsing, heaving thing, with us squirming in the midst of it all.

# 8

# THE GREAT BLENDING

*We are all agreed that your theory is crazy. The question that divides us is whether it is crazy enough to have a chance of being correct.*

—Niels Bohr, speaking to Wolfgang Pauli
at Columbia University, New York, 1958

Perhaps the greatest force of all in the universe is change. Impermanence. Variability. Flux and dynamism, the fact that there is a future. None of us would be here if the cosmos and its contents were fixed and ageless. Yet despite that fundamental debt we often struggle to recognize and acknowledge the kinds of change that matter the most to us, and to life in general.

Earth spins and orbits, and for us the Sun rises and sets and the seasons pass. Organisms reproduce, age, die, and their offspring do it all over again in a ceaseless staccato churn. But our distracted senses and minds tend to inure us to these interleaved cycles, instead blending everything together. You may have never really anticipated the visceral impact of night turning to day at the moment the terminator passes over you, or the orbital instant that summer turns to fall. We're also pretty numb to slow-rolling changes. Yet any adult reading this will have lived through a time when the number of other humans has grown by at least a billion, when global atmospheric carbon dioxide concentrations have increased by at least 10 percent—an enormous shift for a planet—and when species extinction rates, by some measures, are as high as a hundred times above the historical background rate.

At the same time, we have a fascination with the easy narratives of certain kinds of change. Each human generation discovers a technological phenomenon to label as the most disruptive thing the species has yet seen. Back in the 1400s in western Europe it was without a doubt the Gutenberg press, and with some reason. As the German religious reformer Martin Luther is supposed to have stated, in reference to the Protestant Reformation: "Printing is the ultimate gift of God and the greatest one." He should know. He made extremely savvy use in the 1500s of printing to spread his accessible German vernacular translation of the biblical Testaments as he pushed back on Papal doctrine.

Later came steam-powered engines, and we conjured visions of a world filled with boats, trains, and machines puffing away to an imperial future. Then steam was disrupted in the twilight of the 1800s by the lighter and more versatile internal combustion chamber and the rise of a petroleum industry. In the twentieth century we celebrated the emergence of electricity as a transformative commodity, together with the disruptors of telecommunications, computers and the internet, antibiotics, genetics, and of course nuclear energy and weaponry. Here in the twenty-first century we're staring down the multiple barrels of precision genetic engineering, human-induced global warming, machine learning, and the tantalizing possibilities of quantum computing and brain-machine neural interfaces.

But all of these tales really only represent amateur disruptions. They're indicators of something, but in this chapter I'm going to try to explain why I think that we're in the midst of a far, far larger transformation. One that supersedes even global warming, the Anthropocene, and perhaps our own origins a few hundred thousand years ago. It's also not the now-routine kind of proposal for human-machine acceleration, transcendence, or singularity—whatever you want to call it. This is much deeper, and the dataome is at its core.

In order to explain the idea of this transformation we have to take several steps back and take a corrective look at how we characterize the major ups and downs of life's history on Earth. That narrative, the one that most of us have become used to, is all about extinctions and origins in the form of truly massive disruptions and new beginnings. It's a good narrative for sure, but its convenience can also lure us into thinking a little unimaginatively.

Historically this narrative emerged to reconcile two very different viewpoints that lingered on from Western science during the 1800s—those of catastrophism and uniformitarianism. Catastrophism was an idea with religious roots, the notion that the history of life, and Earth itself, would have been largely unchanging were it not for specific, thumpingly violent events, like a Great Flood, or the rapid and miraculous formation of entire mountain ranges. Uniformitarianism, by contrast, was more akin to our modern view of things, where very slow, very incremental changes acting over long times give rise to Earth's landscape and occupants.

Today, of course, we take a combined view. Yes, some dramatic and cataclysmic things have happened, and will happen again, on a very old Earth. But drama is also in the eye of the beholder. And change across ten thousand years is drama for geology and paleontology. Even extinction events often seem to take millions, if not tens of millions of years to play out in their entirety. And the great divisions of time that we've invented, like the four eons and the geological periods—all twenty-two of them, from the Siderian through the Cambrian to the Jurassic and the Quaternary—are defined by complicated and sometimes spotty changes in the rock and fossil record.

The great "Cambrian explosion" of 541 million years ago has been described in such breathless terms that the uninitiated might imagine that one night, all that time ago, life suddenly sprouted new forms and took over the planet in a way it hadn't before. Blobby, often indistinct forms of animals gave way to burrowing, shelled, segmented arthropods, including the famed trilobites, in an explosive emergence of new body plans.

But in truth, recent research and fossil discoveries indicate that this period may have been a time of multiple overlapping branches of life: of species showing up tens of millions of years before the famed "explosion" and lingering on long after, or of pulses of biological innovation that, in the end, left behind only heavily pruned members of those experimental lineages. Behind these pulses? One possibility, as I'll describe shortly, was the shifting oxygen level of the planet. This chemical makeover was a consequence of life for sure, but also of fundamental dynamical feedbacks operating on planetary chemistry and oceanic layering. What happened for life in one part of the world may also have not happened in other parts—creating oases of innovation surrounded by the biological regulars of the time.

A great change like the Cambrian event (and it was certainly great in its totality) would not have felt like an explosion to anything alive at the time. The Sun would rise and set, the seasons would come and go, and generation upon generation would pass without your neighbors suddenly sprouting husky shells and antennae overnight.

A similar patchwork of staggered change applies to many of Earth's extinction events. Long after the Cambrian, sometime around 251 million years ago (back when most of Earth's continental land mass was an enormous and slightly scruffy pole-to-pole Band-Aid), the world experienced the mother of all extinctions. Over a span of a few hundred thousand years, probably in a series of pulses, Earth lost an estimated 96 percent of all marine species and 70 percent of terrestrial vertebrates (including some of the largest insect species ever, members of the griffinfly genus with imposing 27-inch wingspans). This drastic event defines the end of the

Permian period in Earth's history and, eventually, the start of the Triassic period. It was the most extensive loss of species that we think the planet has ever seen. It's known as the Great Dying.

What the precise cause was of all of this is still mysterious. It clearly involved a massive perturbation to Earth's environment that changed the climate, ocean chemistry, and environmental oxygen. But the same shifts didn't happen everywhere, and it didn't take place overnight. Our evidence of the world 251 million years ago is also frustratingly sparse. Earth's own great system of plate tectonics recycles the rocky ocean floor approximately every 200 million years, inconveniently removing much of that record. And what strata there are on our modern continents are mostly deeply buried and tough to access.

Nonetheless, we think that there was a confluence of phenomena behind the Great Dying. We have robust evidence for two major volcanic episodes coincident with the broad timing of the extinction. In what is now Siberia there was an outpouring of lava that eventually coated an astonishing two million square kilometers of land in basalt. That outpouring of planetary guts would have released enormous bursts of carbon dioxide, possibly formed stratospheric ash clouds, spewed toxic compounds, and might have ignited still-forming beds of coal left from earlier periods of exuberant plant growth.

Furthermore, in the window of time shortly preceding the Siberian outflows, the Emeshian Traps in what is now China's Sichuan province had oozed lava across some quarter of a million square kilometers. That episode alone looks to have coincided with one of the earlier pulses of extinction framed within the whole end-Permian event. It is also possible that a massive asteroid impact made a contribution, akin to the one that likely precipitated a mass extinction 66 million years ago, doing away with the large non-avian dinosaurs and relegating the others to a future of diminutive, although successful, featheriness.

Any of these perturbations would have made Earth's biosphere ring like it was hit by a sledgehammer. But while some species were pushed to the brink, other species, particularly microbes, could have bloomed. Some blooms would have cut off oxygen to the oceans and produced masses of compounds like hydrogen sulfide. If hydrogen sulfide got into the atmosphere in quantity it wouldn't just poison plant life; it would deplete the protective ozone layer. To top off the extent of the destruction, there is evidence that ocean temperatures may have gotten as high as 40°C (104° F) in places, turning marine life into a kind of undignified sous-vide bisque.

It was, by human standards, a truly ghastly time. But the Great Dying holds a critical lesson about large-scale disruption. First, imagine if you were alive on Earth in the midst of this event 251 million years ago. Some estimates place the worst of the mass extinction as having taken place in as little as ten thousand to sixty thousand years. In geological terms that is very, very fast, but if you or I were living on Earth in the midst of it all, this timespan would encompass all of our recorded history, as well as much of our prehistory. It's entirely possible that we'd struggle to recognize what was happening all around us. This is an obvious, but often overlooked point. Game-changing disruptions to the order of life on the planet are never going to be easy to judge from the inside.

As much as this Permian-Triassic transition pruned life's lineages, it also resulted in a fresh landscape for surviving species (and genes) to occupy and innovate in. Ten million years after the start of the Great Dying, Earth's landscape for life was like a washed-down café table, mostly empty but for some soggy crumbs. It was ready for the next customers.

Among those customers were so-called "disaster taxa." That name is perhaps unfair—a more generous term is pioneer organisms, the life-forms that survived and could quickly proliferate into largely empty environments. During the Permian-Triassic transition, one of the major disaster taxa living on land were the *Lystrosaurus*, or shovel lizards, a genus of chunky, pig-size herbivores with faces to match. Indeed, they were so successful it appears that Earth was temporarily a Planet of the Shovel Lizards. They were also part of a reptilian group that included the ancestors of mammals.

In the oceans, before the extinction, in the Permian, there had been many non-motile (essentially immobile) species. These were now mostly gone, and instead marine environments experienced a repopulation with scores of free-swimming species. This was a curious, and perhaps difficult upside-down situation for the food chain, still absent its population at the base, likely due to overall low oxygen concentrations in the oceans. It wasn't really until a full 50 million years after the mass extinction that Earth's ecosystems got back up to speed.

But when they did, it was with real style. Because the rest of the Triassic saw the emergence of the first true mammals and the spread and diversification of scores of vertebrates. The washed-down table after the Great Dying set life on Earth onto a new and vibrant trajectory.

The lesson is that real, planet-changing disruptions tend to be complicated, slow-moving, and very hard to spot if you're sitting within them. Even afterward, it's easy to get distracted by appealing narratives of small pieces of the disruption, and fail to see the bigger picture. Critically, there's one further tale from Earth's history that's going to serve as a template for proposing a grander vision of what the dataome represents. Take a deep breath, because that breath is key.

## The Great Oxygenation

Aside from severe mass extinctions or periods of species radiation in life's history on Earth, there are extremely disruptive innovations that have taken major fractions of Earth's present age to unfold. But until each one happened the next couldn't really get started. One perspective on these innovations is that they all involve the unlocking of energy sources. The evolutionary biologist and writer Olivia Judson published a provocative paper in the scientific journal *Nature Ecology and Evolution* in 2017 entitled "The Energy Expansions of Evolution." In this analysis a series of five successive energy "revolutions" divide up the history of life on Earth thus far. These epochs are labeled as geochemical energy, sunlight, oxygen, flesh, and fire.

Geochemical energy relates to an idea that the earliest forms of life were *chemoautrotrophic*: chemical "bottom feeders," drawing energy from basic reactions of minerals and atmosphere. This is not a highly productive route, so the development of *phototrophic* life—utilizing the Sun's photons—represented a boost and an unlocking of a new reserve. Then, oxygen-respiring metabolism unlocked another level of cellular energy use and productivity, as did the "use" of concentrated proteins and fat in the flesh of other multicellular organisms. When the genus *Homo* learned to use fire, it not only enabled better nutrition for itself; it eventually opened up the construction of metal tools, chemistry, industrially produced fertilizers to enhance food and biological energy availability, and so on.

It's an intriguing approach to framing the history of life on Earth. Judson describes it as a way to understand the development of an increasingly complex biosphere on a planet, one that depends on a virtuous circle between evolving life-forms and the changes they bring to the planetary environment itself. If this is accurate, then I think we can add a sixth epoch to the five, and that is the epoch of the external dataome, and of the machine, coming hot on the heels of "fire." The open question is how virtuous the dataome really is, and, as I'll discuss, what fuel reserve it's really unlocked.

The earlier energy expansion that is perhaps most important to us, as a multicellular species, is the transition from a planet largely devoid of atmospheric oxygen to one positively reeking with the stuff. I'll explain more precisely why oxygen mattered so much for biology momentarily. But for now it's sufficient to know that you and I, and every large multicellular organism on Earth, requires freely available oxygen in order to function. The only possible exceptions we know of are teeny tiny marine animals in abyssal depths where oxygen is largely absent. And even these species almost certainly did not originate in these environments.

But 2.45 billion years ago Earth's atmosphere had effectively zero oxygen. Instead, geochemical evidence points to a carbon dioxide–rich blanket around the planet. There may have been atmospheric nitrogen as well (today we wade through 78 percent nitrogen), but there are suspicions that gaseous nitrogen was in decline as microbial life gradually chomped its way through this essential element.

Around that time, 2.45 billion years ago, microbial organisms began to flourish by deploying the fantastic metabolic trick of oxygen-producing photosynthesis. If their numbers had remained modest, then the oxygen that they generated would have mostly just reacted with organic carbon compounds and not seeped much into the environment. But it appears that oxygen concentrations gradually crept up, starting from levels that were less than a one hundred thousandth of those on the modern Earth.

One of the ways we can tell this is by the changes that slowly took place in the compounds now layered into the rock record. Materials like pyrite or uraninite start to disappear from sedimentary rocks at this time, a plausible consequence of chemical reactions with oxygen. There are also distinct layers of iron-rich rocks, so-called banded iron formations (or BIFs). These would be a natural consequence of ancient Earth's iron- and nickel-rich oceans being invaded by oxygen. That oxygen would react with iron to form iron oxides that simply precipitated out to the ocean floors like a vast, gentle marine snowstorm. This kind of planetwide reworking, or terraforming, changed innumerable patterns of chemistry and element transportation around the globe. It was, by any standards, one of the biggest makeovers that Earth has ever seen.

But it wasn't a simple journey from A to B. Starting with those first tentative wisps of oxygen 2.45 billion years ago, it appears to have taken at least another *billion* years before the atmosphere began to really shift its composition in earnest. Eventually, around 850 million years ago, there was a sequence of geologically rapid but unsteady rises in atmospheric oxygen. A notable peak may have been during the Carboniferous period around 300 million years ago, driven by large vascular plants being prolific generators of oxygen, followed by other wiggles in concentration along the way to where we are today.

Somewhere in that longer timespan, perhaps around 600 million years ago, the increasingly oxygen-rich environment enabled the biological innovation of multicellular animals. But as much as the oxygenation of Earth was a trigger for complex life, it was pretty painful in other ways. For lots of microbial organisms then, as now, oxygen is toxic. Also, as oxygen seeped into the atmosphere it would have reacted with compounds like methane, a potent greenhouse gas. Removing methane may have contributed to episodes of a "Snowball Earth," periods when the whole planet dove into a frigid state of self-reinforcing glacial feedback as reflective surface ice exacerbated the dropping temperatures.

The Great Oxygenation of Earth was an enormously long-lived and profound affair. It was arguably the lengthiest revolution the world has yet seen after the revolution of biological life itself.

Other natural disruptions, like extinction events, might result in innovations, but they are not easily predictable or progressive. They are like petulant, violent, unexpected guests. But the rise of oxygen, and life exploiting that oxygen, appears to be nigh on inexorable. A study in 2019 by the scientists Lewis Alcott, Benjamin Mills, and Simon Poulton used computer modeling to demonstrate that once life had invented oxygenic photosynthesis, all the following ups and downs of oxygen's very long history are readily explained by known biogeochemical global cycles involving carbon, oxygen, and phosphorus, with those cycles inevitably taking things in a very particular direction. Oxygen's story is not just predictable, it seems like it's an inherent property of the kind of planet we live on.

I'm going to argue that the dataome represents another slow-burning planetary disruption of a similarly epic scale. Although the chances are that we're currently experiencing one of the more rapid shifts—akin to the Carboniferous oxygen peak 300 million years ago—the rise of data, of information, has been an awfully long time coming. It began with the first replicators on Earth, and the first sensory mechanisms, and it's been gradually intertwining itself with the same biogeochemical processes that oxygen did. Right now it's helping drive a full-scale reshaping of the planet.

## It's Complicated

I've said that after Earth got consistently well oxygenated, it eventually got more complex. A type of cellular architecture for life arose that was the direct ancestor of all complex-celled organisms: the eukaryotes. Modern animals, plants, fungi, and certain groups of single-celled organisms are eukaryotes. Flip through most biology textbooks and you'll see the standard story: First came the prokaryotes, in the form of bacteria and archaea, usually described as simple little bags of DNA and enzymes. Then came the eukaryotes, like an explosion of unprecedented cellular complexity that would propel life on Earth to the vast array of multicellular organisms we know. The eukaryote name comes from blending the ancient Greek terms for "good" (or "true") and "kernel," because a major characteristic is that their genetic material is partitioned into a membrane sack—the cell nucleus.

In these cells are a collective of functional structures above and beyond what most single-celled bacteria or archaea contain (although we'll see that's an oversimplification). Some of those structures are tiny bean-like entities that play a central role in churning out the molecule called adenosine triphosphate, or ATP, which I've referred to earlier in the context of neural energy budgets. In a human, something like $10^{25}$—ten trillion trillion—ATP molecules get deployed per day (being partially broken to release energy via a process called hydrolysis). ATP is like cellular candy—it's a way of dispersing quick chemical energy throughout the cell. Those little bean structures are the mitochondria, and they're experts at taking the chemical products of glucose and a compound called NADH and, together with oxygen, producing ATP. But mitochondria are quite special, because they also contain their own DNA, kept securely separate from the rest of the DNA of the cell.

How this all came to be is a lengthy and still pretty contentious story. The most generally accepted picture is that sometime around two billion years ago the ancestor lineages of complex cells got into a symbiotic relationship with a bacterium-like mitochondrial ancestor. The enormous ATP energy boost this enabled, a factor of 20 or so above what had been possible, was an evolutionary jackpot. Organisms could now exploit the potency of an oxygen-rich atmosphere to power the use of larger genomes and many more genes. And symbiosis turned into *endosymbiosis,* the complete merger of species, and an idea popularized by the brilliant evolutionary biologist Lynn Margulis in 1967. Critically too, mitochondria sit in the complex cells of plants and organisms such as diatoms that employ photosynthesis. Here they can take glucose directly from the photosynthetic parts of a cell and turn it into ATP. In other words, an oxygen-rich world and endosymbiosis also enabled new kinds of organisms to contribute to the flood of oxygen.

The endosymbiosis of mitochondria is quite a commitment. It would be akin to you or I giving over almost all of our bodily functions to someone else, all but for some particularly useful talent, like cooking dinner or fixing the plumbing. Human mitochondrial DNA has become whittled down to a single circular loop of some 16,000 nucleobases, encoding a grand total of 37 genes. It's a lean, mean genetic machine, with all of its other needs outsourced.

In most sexually reproducing species, the mitochondrial genes are only inherited from one parent. For humans the mitochondrial genes typically only come from the egg, and therefore our mothers. Between the stripped-down nature of the mitochondrial DNA and this kind of replication, the genes are remarkably slow-evolving. Some estimates suggest that point mutations, the random changing of a single DNA letter, in mitochondria may only occur once every 8,000 years or so for our *Homo sapiens* lineage. This makes excellent sense in the larger evolutionary picture—you simply don't want such a critical piece of biological machinery changing much from a state that works so very, very well.

But I'd like to flip this perspective around for a moment, because there may be clues to our own future in a world driven by a dataome and its machines. I said that our complex-celled ancestors hit the jackpot when they got together with the ancestors of mitochondria. Arguably it's actually the mitochondria and their genes who hit the absolute mother of all jackpots. For two billion years they've had all of complex-celled life on Earth invested in their survival and stability, keeping them exquisitely preserved. Or, in the language I introduced for the concept of corgs, once the algorithms describing a built-in cellular power station were invented, they have persisted through their instantiation in mitochondria.

But just as we've seen happen with the complexities of extinctions and innovations in the history of life on Earth, the simple narrative of how mitochondria and complex-structured cells came to be the way they are may be too easy. I've mentioned that a major characteristic of eukaryotes is the structures they have inside their cells, known collectively as organelles. These include the mitochondria, but also forms like plastids, vacuoles, and the wonderfully named Golgi apparatus (which acts like a packing and shipping center for proteins). Our received wisdom has been that all of these organelles make eukaryotes vastly more complicated, and therefore sophisticated, than primitive bacteria and archaea.

This picture is now under serious revision. Prokaryotes, in fact, have plenty of internal structures—it's just that they are more species-specific. Some bacteria have membrane-enclosed organelles containing magnetic particles in chains called magnetosomes. These appear to be used for navigation: sensing and exploiting Earth's intrinsic magnetic field structure. Other microbes have organelles called anammoxosomes. These structures seem to undertake energy-producing chemical reactions involving nitrogen.

And there are forms inside bacteria that are not bound inside membranes, but instead are coated in proteins resembling the shell-like coating of viruses that secures their genetic material.

The list goes on, with more and more intriguing structures in the process of being discovered. It's now clear that cellular "complexity" and organelles are not really unique to eukaryotes at all. This raises the possibility that complex-celled life wasn't a sudden innovation. Instead it might have been a slow journey along a path littered with prokaryotes evolving more and more organelle-like complexity (even if very species-specific). The direct ancestors of eukaryotes, which we think may have been related to a modern group of archaea called the Asgard (discovered in the Arctic), would have been a motley crew. The emergence of a well-encapsulated cell nucleus could have just been a late arrival, a capstone to a long journey, and the incorporation of mitochondria would have occurred in parallel with these other trials and experiments. That incorporation was a critical, game-changing step but not an isolated incident.

My point in discussing all of this is to reinforce the observation that there seem to be good examples of "revolutions" in evolution that only appear *in retrospect* to be the result of some momentous and abrupt innovation. In reality they might equally be the result of a slower, piecewise assembly that doesn't even look like a big change until long after the fact. In that sense many innovations may be much more inevitable than they are lucky.

Something else that all of these great changes do have in common is that the innovations they bring seem to stick around. The reworking of the planetary environment from atmospheric oxygenation has not been undone—even though oxygen remains toxic to a large number of single-celled species. The endosymbiosis of mitochondria has not gone away, and, as far as we know, few, if any, complex-celled species have devolved by ejecting their mitochondrial baggage. Even members of the vast group of animals that we often haphazardly label as dinosaurs (they evolved and spread across nearly 200 million years of time) persist today in over

10,000 species of birds. The K-T extinction event 66 million years ago simply did some severe pruning of the lineage.

Having laid all of those stories out, from planetary disruptions to mitochondria to complex cells, the outrageous, crazy question is whether the rise of data and its informational load is not just loosely similar to the oxygenation of Earth, but is actually exactly the same kind of fundamental phenomenon. The sheer scale of the dataome indicates a profound shift in evolution, comparable to the major shifts and events scattered throughout the history of life on Earth.

The closer we look, the more closely the rise of information on Earth seems to parallel the nature of the rise of oxygen. Both involve the reconfiguration of matter and energy flow in very specific ways. The oxygen in our atmosphere didn't just appear out of nowhere. The atoms of oxygen were always here, but now they're reconfigured, and driving even further change to the planet. Information is also a very particular reconfiguration of the world, using all the pieces that were already here or that can be captured, including an endless flood of solar photons and the geochemical mix of a rocky planet. And, like free oxygen, while information has been present for a long time on Earth, its current rate of expansion is exceedingly rapid, and coincident with another special marker that I'll now explain.

## Information Drive

Throughout this book I've followed a line of thought that forces us to increasingly consider information as a "thing," a reified concept. The great energetic burden of human data and the information it contains makes us feel like we're supporting a demanding, albeit often delightful, child. Similarly, the way that information is so deeply integrated into human lives, our individual development, and living systems in general, gives it the feeling of a vital essence. Thanks to people like Claude Shannon, we have placed information within the pantheon of mathematical forms and frameworks that our minds have sculpted from the world. And to top it all off, an information-centric view of life, genes, and evolution has a predictive and interpretive power that's hard to ignore.

Many of us have also gotten used to thinking of the closely related properties of energy and entropy as tangible things. In truth they are really just concepts that help predict or explain why matter behaves the way it does. It's our ingenuity in casting their definitions that allows them to become measurable, empirical entities. Energy, for instance, tells us about the potential for an object or system to *do something*, whether that's an apple falling from a tree and catching Isaac Newton's attention or an atomic nucleus spontaneously splitting off pieces of itself.

But it's hard to point at any phenomenon in nature and say, "That is energy." A photon is not energy; it *has* a specific energy defined by the wavelength of its electromagnetic waveform. It's one reason physicists always wince when a science fiction tale mentions anything being made of "pure energy," because that's just wrong. And don't get me started on the new-age train wreck of energy fields and crystals.

Energy and entropy are conceptual properties of systems in the world, from rocks to photons to the fabric of spacetime. But their mathematical representations demonstrate an elegant utility that compels us to believe that we're onto something, that the language of energy and entropy, the symbolic representation, really is a way of understanding nature directly. And even though thermodynamics at its core unfolds from the statistics of matter's conditions and interactions—the myriad microstates available to atoms and molecules—it is most certainly manifest in macrostates that we can sense and measure; in heat, light, pressure, and the forces and qualities of material substances. These are things that (we assume) evolution has equipped us to have intuition for because they have predictive power.

Shouldn't the concept we call information be added to this rather inbred family for describing natural phenomena? We know that entropy and information are connected. And there are plenty of scientists who already treat information in precisely this way. It's just that we don't yet teach high school physics classes about the laws of information in the way we teach the laws of motion or thermodynamics. Of course, in fairness, this is partly because we don't quite know what those laws are yet.

Our journey through the dataome and the nature of life is helping unearth some answers. Previously I introduced the ideas that researchers are tinkering with to suggest that the *flow* of information is an ingredient in the rich recipe of living things, and that the hierarchical structure of that flow is fundamentally different between living and non-living things. But what if the net flux of that flow is not just an indicator of life but a valid way of measuring an underlying resource, a currency in nature akin to energy and entropy?

Information is a type of rearrangement of the physical world. We've also seen that information can be transformed into available energy. When we boil everything down to the barest of essential physical details, a major piece of what living things do is move electrical charges around. This is true on the scale of individual microbes all the way up to the planetary scale, across the entire biosphere. Organisms, through their chemistry, are endlessly shuffling electrons and ions from one place to another. In doing so they can repurpose energy and make things happen, from building new molecules to physically repositioning themselves.

Tracking those movements of electrical charge tells us lots about the functional nature of life. We learn something about the flux of information this way because information can be carried by electrically charged objects, be they electrons, ions, or molecules. But information can be carried in other ways too. It's not so fussy about how it's manifested in the world.

Information movement and generation is also about computation in its broadest sense: to the carrying out of chemical reactions or calculations whose outcome can provoke new selections or trajectories. Like our imaginary Darwinian demons who are busy computing the choices that best propagate living things in the world, for organisms there are, in effect, computed solutions to overcoming barriers to movement, change, and growth—more reconfigurations of matter. These computations find the pathways to changing the state of a system, like switching a lightbulb on or off. On a much grander scale, in my conversations with various researchers I've frequently heard the following bird's-eye description of what's happening all around us: *Life on Earth is nothing more than a four-billion-year-old catalytic chemical computation.*

What, you might ask, is it computing? The answer is that it's computing its own future. Life is, as we talked about earlier, the ultimate algorithm. At the core of the computation is information manipulating information, just as a piece of software can manipulate another piece of software or itself. But as we saw, information can be highly contextual. For life it would seem a fair bet that the most meaningful information is that which influences survival by enhancing fitness. That meaningful information is the truly vital resource, running in concert with the flow and shuffle of electrons and ions. It is the fuel that drives all of us.

I can't raise this picture without stepping back for a moment from these profound concepts and recalling that the idea of our planet as one big, funky computer has come up before. But unlike the fictional computer in Douglas Adams's brilliant 1979 novel *The Hitchhiker's Guide to the Galaxy*—where Earth is a product of an alien race that wants to know what "the ultimate question is" if the *answer* to life, the universe, and everything is "42"—our scientific vision of a planetary computer is about chemistry, energy, and information computing the solutions to propagating itself into the future. In that sense, the answer to life, the universe, and everything isn't 42; it's survival. And the ultimate question is what meaningful information is necessary to make that happen.

We can even gauge the scale of this planetary computation. Our best estimate of the number of DNA base pairs (the opposing parts of the rungs in the ladder-like double helix of the DNA molecule, and the fundamental encoders of biological information) in *all* of extant life on Earth is around 50 trillion trillion trillion (or $5 \times 10^{37}$).

Taking this number of base pairs, it's possible to extrapolate to produce an estimate of the total computational "processing rate" of the biosphere. This was attempted by the researchers Hanna Landenmark, Duncan Forgan, and Charles Cockell in 2015. In this case the fundamental unit of computation is thought of as whenever a nucleotide (a base in the base pair) exerts itself chemically. It may be sensed by a transcription process—turning a stretch of DNA into messenger RNA—or replicated during cell division. In computer science we talk about floating-point operations. That refers to arithmetical steps like adding or subtracting two numbers with decimal places in them. The speed of a processor or system is then labeled in terms of floating-point operations per second, or FLOPS. For biology we can talk about nucleotide operations per second, or NOPS.

By this measure the modern Earth has a net biological processing power of about a trillion trillion ($10^{24}$) NOPS. When it comes to machines, some of the largest supercomputers we've built manage around $10^{17}$ FLOPS. Although a direct comparison is a little fraught, we might claim that the total terrestrial biocomputer is something like $10^{7}$ times more powerful. Even if we add up all the human-made computing power that exists today, it clocks in at around $10^{21}$ to $10^{22}$ FLOPS. That's if every single processor on the planet was running at full capacity at the same time, from supercomputers to the chips in your washing machine or your musical toilet.

These numbers seem to be telling an intriguing, albeit complicated, story. The first observation to make is that while the NOPS total for Earth is still larger than the total FLOPS, it's only off by a few of orders of magnitude. Another observation is that NOPS and FLOPS clearly represent fundamentally different kinds of processing. A nucleotide and its base pair partner are arguably far more information-rich than a handful of bits in a silicon-based processor. Single base-pair influences can change the outcomes for individual organisms and for the distributed populations of a species' lineage across millions of years of time. The current NOPS level of the planet is just sampling a computational system running algorithms that span geological timescales.

Yet the gap between raw NOPS and FLOPS is definitely going to close in the near future. As I've shown previously, the recent rate of growth in electronic computing capacity is somewhere between 50 percent and 80 percent each year. That implies that we might reach parity between raw nucleotide processing and artificial processing within the next ten years. This is particularly likely as we expand our internet of things, embedding more processing power into everything from cars to toothbrushes.

While the world is unlikely to abruptly transform when this parity is reached, it is most certainly a special milestone. Never before in the four-billion-year history of life on Earth will there have been an equal, or superior, system of information processing to that of living organic chemistry. From the perspective of disruption, of evolutionary innovation, and of origin events, this will represent a fundamental marker and transition—the kind of transition that we scrabble around in rocks to see the signs of in fossils and geological tracers.

That it coincides with an ongoing extinction event—or at least substantial evidence for such an event in the depletion of species and rapid changes in planetary climate—just adds to the significance. The parallels with a change like the Great Oxygenation of Earth's atmosphere are compelling. Although that shift played out over much longer timescales, and with more variance, that's a fact only seen in retrospect. As I've argued, seen from inside, massive disruptions can be hard to recognize.

Imagine for a moment that the rise of the dataome, and the transition to non-biological computation on Earth, is more like the very first burst of molecular oxygen that really persisted some two billion years ago. The human dataome is the first burst of non-biologically-held information to hang around in the environment. It represents a gateway to greater and greater machine dominance in Earth's ecosystems and the release (for want of a better term) of more and more new information into the environment. It represents the cornerstone of a fundamental shift in the planetary environment. It represents an origin event.

A threshold has been inched across over the past two hundred thousand years to what is potentially millions, even hundreds of millions, of years of a future we cannot yet see. Like the first measurable oxygenation of Earth, it comes with extinctions, the pruning of lineages that can't cope with the new environment. Oxygen was toxic to much of the life on a younger Earth. Perhaps the rise of the dataome and information also has a natural toxicity to many species. The only difference is that the toxicity is more indirect, and is felt through the changes forced on the environment in support of the datatome and the metalworld. It is unclear whether or not that toxicity extends to us too.

If we're sitting inside this major evolutionary event as it plays out, with both extinction and origin in synchronicity, what does this mean for us and subsequent generations of humans? Let me propose a further, outrageous but strangely hopeful hypothesis about us and our future.

## Carry Forward

In these past two hundred thousand years, humans have incubated a dataome. That dataome has, as evidenced across the pages of this book, become increasingly intertwined with our survival and our behavior as a species. By several measures, including the comparison of global NOPS and FLOPs capacities, and the rise of our metalworld's resource demands, it's not only bigger than us, it's on track to eventually being bigger than Earth's biological life.

Furthermore, from an evolutionary perspective, I've argued that humans and the dataome may already constitute a holobiont. Such an entity may experience selection forces that act on its emergent properties—on the song not the singer. Those forces operate on the net outcomes of humans, data, and machines all mushed together. The steps beyond simple symbiosis may already be happening. We are on a trajectory toward a new endosymbiosis, the kind of extraordinary blending of species that likely produced our complex-celled ancestors in the first place.

But what if, instead of humans absorbing or melding with machines, we are both better served by humans being the absorbed party? In biology the mitochondria generate ATP, essential chemical energy for all complex cells. For the dataome humans generate the *one thing* that we have yet to see machines or artificial algorithms produce: original information, real innovation, and open-ended novelty. Ideas seemingly plucked from the ether.

This might sound like another Faustian bargain, wherein we have to give up some supremely important human essence. But suppose that's not the case. Suppose it is *precisely* the things most precious to us—our imaginations, our seemingly boundless creativity—that are an essential currency. We inject new, meaningful information into the world. In return (if our story is going to parallel the incredible evolutionary success story of mitochondria) we get to be preserved and maintained, largely free of the inequities and tyrannies of future natural selection. Forever human.

And the really important observation about all of this change is not even that it's accelerating right now, or that it picked up pace a few centuries back with innovations like movable type. The most critical point is that in truth we have *always been* a part of something new. When we speculate about human transcendence, or technological singularities, or post-human futures, we're missing what's right in front of us. Evolution has been tinkering with ways to better propagate information on Earth since the get-go four billion years in the past. Other species have had diminutive, modest versions of a dataome. But a few hundred thousand years ago the emergence of these creatures we call hominins cracked open the door to a growth in this pattern. Languages of abstraction and symbolism, ocher handprints, tools, and histories all started gushing forth.

*Homo sapiens* got lucky and kept on going, coevolving with its external data. An oddity became an evolutionary success story.

In science we often struggle philosophically with the notion that there is something special about humans, something unique. Yet you just have to look around the world to wonder if we aren't actually very strange. The answer could easily be that yes, we are strange. We're the manifestation of a breakthrough moment in billions of years of natural selection, an informational experiment that finally clicked into place. That doesn't make us the last chapter in the extraordinary story of life on Earth, but we're definitely the archetypes of what comes next.

# 9

# A UNIVERSE OF DATAOMES

*We die. That may be the meaning of life. But we do language. That may be the measure of our lives.*

—Toni Morrison, 1993, Nobel Prize in Literature acceptance speech

Throughout the information-infused pages of this book I've examined human existence through an unusual lens. That lens is the dataome, and its peculiar optics bring into focus some of the most complex and unsettling stories we have thus far accumulated about the nature of our species, and of life itself.

The sheer weight of our external data is astounding, and in the context of how we exploit planetary resources, worrying in the extreme. By some appearances we've triggered a self-amplifying, catalytic set of planetary changes, with the dataome arguably sitting at the nexus of most, if not all, of our behavior. I've also suggested that there is little reason to think that the dataome is perfectly healthy. It could be quite dysfunctional in parts. That's not to say that it's destined for extinction, because like any dynamic and evolvable phenomenon it will swing back and forth, testing out the valleys and hills of a vast and complex landscape of rules and opportunities.

But that landscape certainly has hazards. Ironically, some of the same scientific principles that we've discovered about the unexpectedly fundamental nature of information and computation are themselves contributing to the surge in the growth and demands of human data. At the same time, these principles and the tools they spawn—from machine learning to blockchain—could offer hope for an improved stewardship of the dataome and the planet. Yet as we chase the deeper meaning of information, and the implications for our future, we come up against a disturbing possibility. Not only may we be inseparable from our dataome, we may be a component—an organelle or part of a holobiont—in a genus (beyond merely a species) that represents a massive, slow-rolling evolutionary revolution on this planet.

That revolution involves the unfettered, amplified release of information in the way that photosynthetic life pumped oxygen into Earth's environment. We humans could be the informational equivalent of the mitochondria inside complex cells, a set of endosymbionts preserved but also trapped. Our era of transcendence is already well behind us. It happened in the gradual changes that took place as we speciated from our fellow hominins somewhere in sub-Saharan Africa all those hundreds of thousands of years ago; with our incessant talking and making-of-things.

If all of that were not strange enough, the phenomenon of life itself can be conceptualized as nearly divorced from its physical elements: simply a representation of self-propagating, flowing information, governed by principles we're only just starting to grapple with. Indeed, in a layer beyond even genes and organisms it may make sense to see the world as an emerging, hierarchical collective of meta-function, of core algorithms. The ideas for seeing, eating, flying, and just being, all forcing themselves downward through humble matter to cause it to snap to attention, to get up and walk and talk and create.

As challenging as these perspectives are, we can't stop there. Because all of these explorations, all the stories in this book, from Shakespeare to

Shannon, cuneiform to khipu, punch cards to data centers and metal-worlds, are linked in a way that suggests why all of this has happened to us, and why it will have happened again and again across the universe.

## Energy, Entropy, Information

Let's go back to the business of energy. Parochial observations of our world tell us that the measure we dub "energy" is neither created nor destroyed, and that all material things, whether atoms or elephants, "seek" both their lowest energy state and to be in equilibrium with their surroundings. What do those things mean?

If there is a way for something to shed its energy then that something is not in its lowest energy state. Imagine that you lean too far over on your bicycle. As you crash to the pavement you are releasing, or dissipating, energy to the rest of the world. As you lie there feeling bruised, you realize that at least there is nowhere further to fall. You're literally in your ground state of lowest energy. That state is a condition where there is nothing left to give up, nothing that can be extracted from you; nothing more for you to do to change the rest of the cosmos—at least in a localized, relative sense (your motion on a spinning, orbiting planet and your body's chemical bonds are both signs of not truly being in a cosmic lowest energy state). But there can be an enormous range of how likely it is for something to shed its energy and drop down to that lowest possible energy state. You can think of that range as how probable it is for you to tumble off your bicycle.

Take a cup of hot coffee as another example. You might set this down to cool off for a few minutes. During that time molecules of water and coffee-related chemicals will evaporate into the air, carrying a certain amount of thermal energy away. That lowers the temperature (which is simply a statistical measure, a macrostate, of the thermal jostling of every atom in the liquid). The coffee will also be conducting thermal energy into the cup, whether it's porcelain, glass, or paper. From there, and from the liquid itself, there is a glow of infrared light, of photons produced as molecules bounce, jostle, vibrate, rotate, and occasionally squirt some of that energy out when they fall into slightly lower energy configurations. Because there are septillions ($10^{24}$) of molecules in your cup, it is a complex system and cannot cool instantaneously. But there are other things in nature that can find their lowest energy much, much more quickly.

A subatomic particle called the muon, a heavy version of an electron, has an average lifetime as an isolated entity of around 2.2 millionths of a second. If a muon is created by some process it will sit around on average for this length of time before shedding energy. It usually does this by decaying into three new particles: an electron and two neutrinos carrying energy away to the universe. But the electron and the neutrinos are, if left alone, not going to become anything new. Compared with your cup of cooling coffee this is a more sophisticated business, involving the transmutation of matter. But the underlying impetus, the dissipation of excess energy, the shift toward a state of future inaction, is really just the same.

In the previous chapter, we saw how one of the most useful ways we've discovered to think about the nature of life and its underlying chemistry is in precisely these terms: that global catalytic chemical computation (life on Earth as an information-crunching process) is all about survival, but it's also all about energy transferal, energy dissipation, and the impetus toward an equilibrium with the rest of the universe. And that principle applies across all scales.

We can think of the entire Earth as an object spending its time trying to dissipate energy and return to being an inert smear of atoms adrift in the cosmos. Earth is constantly sliding toward a condition of inaction and greater entropy, one of the possible futures that vastly outnumber those of order and instability. But it has to contend with the ceaseless input of electromagnetic energy from the Sun: heating Earth's atmosphere, oceans and continents, and triggering chemical reactions. Our rocky planet also has interior energy sources, dominated by the radioactive decay of elements like uranium and thorium, held for billions of years inside what is in effect a heavily self-insulated sphere of mineral gunk.

Most of what energy Earth sheds to the cosmos is in a high-entropy infrared glow. It's like that cup of hot coffee, on a much grander scale. All of the input energy cascades downward to what we call thermal energy—the jostling agitation of atoms and molecules that we associate with the property of temperature. For instance, a set of visible solar photons may be neatly absorbed by a plant's photosynthetic molecular machinery. The energy they carry is transferred into the motion of electrons and the promotion of chemistry, resulting in the release of oxygen and the formation of things like sugar molecules. But none of this is a hundred percent efficient, so some energy winds up as waste heat. That heat has high entropy, meaning it is, well, kind of useless, and ultimately it only contributes to the temperature of the planetary surface and atmosphere before radiating away to space. So, our planet sits like a beacon of infrared light in the cosmos, a tiny glowing ember of dissipation.

All of life participates in this. Part of everything that you do, from eating food to running a marathon, clapping your hands, or reading some Shakespeare, will eventually make its way into the thermal glow of Earth. Today there are energetic remnants of all the humans that came before you, only now making their way out to the universe in infrared photons. There's a photon whose existence was set in motion by the efforts of a cuneiform scribe. There's another whose lineage can be traced to when a warlord asked for a new steel sword to be cast and formed. There's one from the plow that tilled the soil that grew the wheat that provided one meal on one specific day for a young physicist called Albert Einstein. All of us, and all living things, have an unavoidable role in the grander dissipation of energy on the planet.

But that's not quite the whole story. The thing about energy dissipation is that it is like a cascade of water down a hillside. It is constantly seeking out paths to follow. Some of those paths are shortcuts, ravines or tunnels. Many paths are ordinary, just allowing individually modest trickles and rivulets. Other paths require the accumulation of water in a pool before filling it up to a critical level where there is a sudden release—a renewed flood down the slope and across obstacles.

Life appears to relate to some of these shortcuts or temporary pools that unlock new routes. The chemistry of life, as we've seen in a general sense, is a set of catalytic reactions—reactions that trigger or enhance other reactions. In fact, it's even more than that. The chemistry of life is auto-catalytic—it amplifies itself. The physicist Eric Smith, whom we encountered before, has described how we can look at the conditions of the young Earth, some four billion years ago, as a landscape of steadily increasing chemical pressure or potential, a bit like those filling ponds on a hillside. Life, in this picture, was an almost inevitable mechanism to help release all of that pent-up potential. Life unlocked new routes to let the metaphorical flood carry on. The complicated chemical reactions of living things let energy continue its journey of dissipation out to the cosmos.

But, as we've seen, living things are also examples of the rare low-entropy fluctuations that can deviate from global equilibrium states. These far-out deviations are precisely where the most interesting, novel, and open-ended structures form and evolve. Think of the cascading innovation that happens in every pot of water that you heat, where convective blobs of liquid rise and fall in a patterned-yet-unpredictable way.

The cell of a microbe or animal is a wild party of non-equilibrium processes. It snatches energy from its environment; it carries out chemistry and one-way gene transcriptions that are far from a statistical balance. It communicates with other cells, and sometimes it can propel itself with teeny molecular motors. Critically, though, if we step back far enough we can see that the net result of all of these deviations is still the same. There is still a net dissipation and an inexorable drift toward a future equilibrium. It's just that being temporarily out of balance can actually get you there a little bit faster. The same kind of principle is at play over microseconds of time in microscopic organisms as it is in thousands of years of deep ocean currents and atmospheric wind patterns. All help move energy around, and dissipate it toward a grander, cosmic equilibrium state.

Relatedly, the physicist Jeremy England, in 2013, put forward a theory for the origins of life as, in effect, a thermodynamic imperative. In the right conditions this imperative causes matter to restructure itself (what England calls dissipation-driven adaptation), in service of energy loss and the growth of entropy. The basic fact of the emergence of life's information flow and function may, in the end, have entirely to do with this kind of known physics, even if the phenomenon of life in its entirety is not easily corralled into a fundamental theoretical framework.

What does all of this have to do with the dataome? Think about some of the waypoints we've visited in this lengthy journey. Recall the fact that building devices like microprocessors or memory chips involves the creation of low-entropy structures at considerable energetic cost. To utilize these devices, to store and manipulate data and its information, also involves the creation of highly organized structures, whether in electrons or as magnetic blips in hard drives, and whether permanent or temporary. The rest of the dataome may be less extreme, but it is not really any different. Hand stencils on caves, imprints in clay tablets, knotted khipu, punch cards, and billions of printed pages are all, ultimately, rearrangements of matter into more-organized structures.

All of this involves the transformation of energy, and inevitably the loss of useful energy into higher-entropy forms to offset these exquisitely organized constructions. It is true that we have gotten more and more efficient, driven by our own short-term needs and goals. It's far more efficient to print a book today than it was four centuries ago. And silicon-based processors are vastly more efficient in terms of computations per unit of energy expenditure than they once were. But, as I've shown, we are also constantly growing our demands. So, the upshot is more and more energy transformation, and more and more waste heat—energy given over to thermalization of the planet. We might be individually more efficient, but there are many more of us doing many more things.

The complicated balance between costs and benefits in energy use reveals itself as driven by the same principles. Some examples reach down to the fundamentals of organism behavior and evolutionary fitness. A beautiful study published in 2018 by Marius Somveille, Ana Rodrigues, and Andrea Manica asked why there are so many different migration patterns for bird species around the world. Some birds barely flap across a few hundred feet to go between their day-to-day habitat and a breeding ground. But 15 percent of avian species redistribute themselves across the planet on a seasonal basis.

An extreme example of that migration comes from the modest-looking arctic tern. This 100-gram aeronaut of feather and flesh travels as far as 60,000 miles in a meandering, indirect round trip between the North Atlantic and Antarctica every year. The researchers used computer simulations to demonstrate that the reason for such an extraordinary journey could simply be one of thermodynamics. A bird species needs to optimize the balance between energy demand and supply in an environment full of competition. For some species, to shift from cold high latitudes to warm equatorial ones, and on to cold again, actually makes sense from this perspective, even if it involves such vast distances. But optimizing the energy balances this way clearly also involves a lot more overall energy expenditure, and therefore much more opportunity to participate in the grand dissipation of the planet. While it's doubtful the arctic tern knows this, it is fully committed to something much bigger than itself.

Seen from the high plateau of physics, all of what birds and humans are doing makes absolute sense. With the dataome we humans are, in our way, doing *precisely* what we should in order to increase the flow of energy, the glow of infrared photons away from Earth and out into the universe. It doesn't matter to physics that in doing so we change our planetary environment, making it less and less habitable for us as a species. A catalytic phenomenon runs until it can run no more. Neither physics nor evolution can really see the future, nor do they care.

The rise of the dataome, of an information symbiont to biological life, may be just another example of a pressure release mechanism for the pent-up energy potential on the planet. It could still include something else too: a reflection of physics that extends far beyond our classical ideas of energy and dissipation. That possibility is the most exciting of all.

## Bits All the Way

Early on in this book I mentioned the physicist John Wheeler's concept of "it from bit": the notion that the properties that we ascribe to physical reality are only evident because they are encoded by information, by observers querying the world. To quote Wheeler himself (from a lively 1989 paper in which he tackles the question after framing it as simply, "How come existence?"):

> It from bit symbolizes the idea that every item of the physical world has at bottom—at a very deep bottom, in most instances—an immaterial source and explanation; that what we call reality arises in the last analysis from the posing of yes-no questions and the registering of equipment-evoked responses; in short, that all things physical are information-theoretic in origin and this is a participatory universe.

The truly fascinating thing is that in our efforts to further enable the dataome, to increase our computational efficiency and storage capacities, we inevitably turn to more and more fundamental, and tiny, parts of the physical world.

It was the physicist Richard Feynman, in a now-famous lecture in 1959, who used the phrase "There's Plenty of Room at the Bottom" in his talk title. What he was referring to was the (then largely hypothetical) manipulation of matter at an atomic level to explore and exploit chemistry, as well as to make machines and other devices of inconceivably small scale and novelty. What Feynman had recognized was that by building and manipulating at the microscopic level you can create vast terrains of complexity in a way that the macroscopic world simply won't allow.

Today we have nanotechnology, and we use devices like atomic force microscopes that can be exploited to sense and manipulate individual atoms. Researchers can readily encode information in DNA, and utilize biochemical machinery to perform actual computations at a molecular level. We also have the beginnings of quantum computing. Although I've expressed skepticism about the hyperbole that presently surrounds quantum computing, that doesn't mean that the potential isn't extraordinary. To the contrary, in many respects the idea of using nature's fundamental qualities to process data is vastly more sensible than anything we've done thus far with silicon chips.

The mathematical underpinnings of quantum computation unfold from a comparatively simple (well, "simple" may be a little optimistic) generalization of our theory of probabilities, involving an upgrade from the "bit" and classical Boolean logic to quantum, or qubit, logic. The physics of quantum superposition, entanglement, and interference—all necessary for quantum computing—are entirely natural. In principle quantum computing doesn't require the tomfoolery to coerce electrons to move around conventional semiconductor logic circuits. Those highly engineered material compositions, involving forced "dopings" of substances with traces of other elements, are an awfully messy way to try to squeeze out just the right kind of on-or-off behavior in an electrical circuit. Instead, in quantum computing the great challenges are in creating the necessary environments for quantum logic to be able to play out, and for it to be successfully read out through the inevitable noise of a system. There's also the challenge of actually designing quantum algorithms. But the actual inner workings of quantum computing and logic are built into the universe, ready to go.

All this means that, from nanotechnology to quantum computing, we can or are at least pretty close to being able to encode and manipulate information using the purest substrates of physical reality, whether electrons or atoms, photons or molecules, or macroscopic quantum objects. Of course, we're not the first. Life itself is already encoded in these base levels, in RNA and DNA, and the vast families of proteins that build all organisms on Earth. Chemistry is a quantum process. And there is some, albeit controversial, evidence that spooky quantum phenomena—like entanglement—may play a role in certain biochemical processes. For example, the dominant mechanisms of photosynthesis used by life on Earth may gain an efficiency boost through the existence of quantum entanglement and coherence over very short distances.

But all told, with our modern implementations of the dataome we're making a new contribution to the specificity of order at the smallest scales in the universe. In effect, our minds are rearranging reality in service of information meaningful to us, with an organizational logic that in many cases is stricter and more rigid than that of biological information. What this also means is that, in humans, the cosmos has hit upon an experiment that unlocks some of the deepest levels in the game of energy dissipation. Part of the trick, the special key, is that our informational selves—represented by the dataome—involve an extreme local reversal of the second law of thermodynamics. In a sense we drill backward into a growing pile of entropy, forging our low-entropy reservoirs of data and information. By the time we're manipulating qubits in service of computations, we've gone as far as we perhaps can.

All of this indicates that we are beginning to close the circle on John Wheeler's "it from bit." Thinking about this at a very superficial level: when we encode information into the configuration of atoms or quantum mechanically entangled objects, we are making structures with very specific physical properties that subsequently engage with the rest of the universe regardless of our intended informational content.

For instance, a designer molecule that encodes a few bits of data in some hypothetical molecular storage system shows two outward faces. For us there is the encoded data, the elements and molecular bonds in some sequence. But for the cosmos there is a structure that has the potential for unique chemical reactions and physical attributes. The fact that it was our information that resulted in this specific molecular structure seems irrelevant to the rest of physical reality, yet without those bits this molecular "it" wouldn't exist.

A more everyday example is when you or I tap a key or a touchscreen. Each letter I write here is being encoded in tiny electrical charges in a piece of engineered silicon. Electrons are being positioned in one microscopic piece of the volume of the entire universe because of my ideas, my creation of information from my wet, warm, intricate brain. Their arrangement, their very existence, is because of information. I think this is true whether or not one buys into the concept of downward emergence that we came across before. Equally, in my brain are neural connections, electrochemical changes, and neurotransmitter molecules that are both the result of information and, somehow, the originators of the information that has now led to those electrons' placement. We are literally it from bit.

Of course, at a much deeper level Wheeler's suggestion was that what we consider as reality is in truth a manifestation of the interrogation of phenomena (or a manifestation of their interactions). Ultimately everything boils down to yes or no, 1 or 0 questions and answers, and only becomes real as a result. His "negative twenty questions" thought experiment, where the answer is never decided upon until the final question, illustrates this. The real motivation for Wheeler's vision comes from the idea of observer-participation in quantum mechanics. This is the idea that "observations" (meaning interactions in order to make measurements) cause otherwise undefined, fuzzy superpositions of quantum wave functions to snap to attention, to manifest things like definable subatomic particles. Observations unavoidably influence the system being observed. Taken to its extreme this suggests that the very phenomenon of conscious awareness is what makes reality. Which is, of course, an excellent topic for pot-smoking students of all ages.

Personally, I'm not sure a sentient mind is so critical. The same interrogation occurs when, for instance, two atoms "decide" to form a chemical bond or not. A probabilistically determined event is still a form of decision between options, and therefore involves information or bits. Nonetheless, by instantiating our information in things like qubits, and even classical semiconductors, we sentient beings are causing parts of the universe to snap to attention, to construct reality in a way that would not have happened otherwise. In ways we can't quite see yet, we may indeed be rewriting the rules of physics.

There is an even more fantastical-sounding idea that sits somewhere among these concepts of information and the reality of the physical world. That is, to use its common label, the *simulation hypothesis*. In this hypothesis it is posited that if a species in the universe ever becomes capable of simulating self-consistent realities (entire universes)—perhaps through advanced, generalized quantum computing—it is more probable that we're in one of those simulations ourselves than in a "real reality."

One argument for this comes from the Swedish-born Oxford philosopher Nick Bostrom, who wrote about it in a 2003 paper. In a nutshell, he suggests that if such simulations are possible, and if a species progresses to a "posthuman" level of technological prowess, they might be interested in running simulations of their ancestors and the evolutionary history of life. Computers this powerful would also be capable of running many, many such simulations. Therefore, it would be likely that most sentient individuals are actually simulated, far outnumbering those in the original physical universe. Rationally then, it's far more probable to find yourself to be a simulated mind than an original one, which is either incredibly exciting or utterly soul-destroying.

I think it's an astonishingly high bar for us to buy into the idea that we're all simulated (and to be clear, Bostrom is much more level-headed about the possibility than many who beat this particular gong). Yet I also think that it has a logical connectivity to the dataome and with all of the proposals I've made, including core algorithms, our mitochondria-like future, and the ever-deeper manifestation of our information in the physical substrates of the world.

To run simulations of entire realities is theoretically plausible, inasmuch as quantum qubits and their entanglement offer the exponential scaling required (in principle, around 300 perfectly operating qubits could allow us to process all the information representing all particles and photons in the universe since the Big Bang, according to the MIT physicist Seth Lloyd). How long those simulations would take is another question, as is how you'd enter the information of the initial conditions (setting up a universe is a lot of work). Or, indeed, how you'd interrogate and analyze their output. Nonetheless, it would be the ultimate expression of the dataome to simulate a reality, presumably containing organisms busy generating and propagating their own dataome. That virtual dataome would dissipate virtual energy that, necessarily, results in the dissipation of real energy; all that virtual energy must be represented by information-carrying phenomena in the simulator. It's an intriguing illustration of the fact that even software has to be instantiated in *something*.

Irrespective of the details or fantasies, the bottom line is that the more we implement information and computation in the fundamental substrates of the physical world, the closer we get to what I'd call a virtuous cycle for thermodynamics. In such a cycle, the configuration and reconfiguration of matter utilizes all possible pathways for dissipating energy, for racing as fast as possible to a place where, somewhat ironically, there is nothing more to happen in the universe. And that raises some questions about what's going on elsewhere in the cosmos, and what will happen in the far, far future.

## Cosmic Imperatives and Endpoints

Unless you've been living under a rock for the past three decades, you'll know that astronomers have been on a planet-discovering bender. Thanks to technological advances and sheer persistence, we now know that statistically speaking most stars in our galaxy are hosts to planets, which we collectively label as exoplanets. The universe need not have been this way, but it seems that other worlds are abundant and that the odds are high that many (possibly billions in our Milky Way) are rocky, damp, temperate places not unlike our Earth.

What we do not know, for now at least, is whether any other places in the cosmos have initiated and sustained living things. It's both an age-old question and at the absolute frontier of modern science, much like the efforts of ALife that we sampled. As with many big and important questions, there tends to be a bifurcation in opinions about the answer. Most scientists are either for life being ubiquitous or against it, with the latter stance being the most convenient for explaining why we've neither spotted signs of distant biospheres nor heard the dulcet tones of alien technology ringing out across the cosmos.

But, in a universe filled with planets (and other nice incubators, like moons), and where the chemical ingredients for life as we know it (made from carbon, nitrogen, phosphorus, sulfur, hydrogen, and oxygen) are absolutely everywhere we look, the notion of life being extraordinarily rare almost takes us back to the ages-old notion of vitalism. That is the conceit that living things are governed by different principles than anything else in the world; that it takes a "vital spark" of something near-magical for life to occur. Personally, I don't buy this for a moment, for two main reasons. The first is that, while we've seen repeatedly that life is a ridiculously complicated thing, that statement says as much about our human limitations as about the phenomenon itself. Just because we don't fully understand something, that doesn't mean it exists outside our known parameters. The second reason relates directly to what I've described as the thermodynamic imperatives that appear to lurk behind living systems: energy dissipation, entropy growth, and the persistence of informational entities simply because they can persist are all irresistible, and recognizable drivers. Admittedly, neither of these reasons tells us for sure that life starts up often, but they remind us that it need not be rare, or compelled into existence by anything mysterious.

This "life happens" viewpoint takes us directly to the question of whether there are also numerous examples of dataomes out there in the universe. If we're comfortable with the notion that biology akin to that on Earth frequently pops up where it can, and that some places will eventually converge onto the algorithm of technologically inclined intelligence (the core algorithm for language and symbolic representation, if you will), I think dataomes are an inevitability.

The existence of other dataomes would improve our odds of identifying other sentient species in the cosmos, because dataomes are tied to technology—the intentional repurposing of matter and energy. Technological signatures (technosignatures) might come from listening for radio signals, looking for powerful laser pulses, or by stumbling across signs of any kind of agency-driven rearrangement of the world. Those rearrangements could be in the form of large off-world structures, space engineering, or even industrial pollution. Behind any smart biological species there would be a co-joined dataome, busy bending everything around it in service of information.

That raises some intriguing possibilities and questions. For instance, I've argued that humans are experiencing a special period of transition. Our dataome is growing so fast right now that its impact is felt across not just human society but the planetary environment as well. It's also on the cusp of reaching, and surpassing, parity with the "catalytic chemical computer" of life on Earth (even if that comparison is complicated).

We don't know what is yet to come for us, but out there in the rest of our galaxy there may be places where all of this has already happened—where there are dataomes, or biology-and-dataome holobionts, that are hundreds, thousands, millions, or conceivably billions of years further along their evolutionary trajectories. If those dataomes have blended endosymbiotically with their biological originators in the way I've speculated is possible, perhaps life in those places is mostly dataome and just a speck of biology. Dataomes and their associated metalworlds are all that the casual observer would ever see.

Or, given the essential role that biological life plays in generating novelty and surprise—perhaps as a mitochondrial-like piece of things—it's not hard to imagine places where machines and biology still fully coexist but are inverted. We've seen that the placement of information storage for evolvable systems need not be in individuals per se. For machines that's certainly true. And for biology on Earth it has long been the case that functionally critical genes are held in a distributed fashion across species; a kind of ultimate backup system for information that has to do with critical metabolism or body plans.

What if there are alien dataomes that have subsumed all of the information of their biological ancestors and now play the role of Darwinian demons? Worlds where the algorithmic cores of DNA are held in nonbiological storage, modeled, simulated, and then used to construct bona fide biological organisms when needed? Self-replicating cells and creatures could be 3D-printed on demand according to the decisions and needs of machine life. You want a nice bespoke forest ecosystem? No problem, we'll print out the trees, birds, mice, fungi, soil bacteria, and viruses and set it up for you. You want to simulate a billion years of possible evolution and then build those creatures? Sure, just let us know how many and whether delivery next week is acceptable.

There are other similarly exotic possibilities. A mere 40,000 years ago *Homo sapiens* still shared the planet with *Homo neanderthalensis*. What if these two lineages of sentient, intelligent species had continued side by side? Conceivably they might have generated a shared dataome, a nonbiological interface for the minds and behavior of different species. Indeed, we might speculate that this did actually exist, in rudimentary form, while *Homo sapiens* coexisted with other hominins. In a less direct fashion today we do actually share parts of our dataome with other species, it's just that those species are not actively contributing to the dataome in the way we do. Every pet photograph or zoological social media group, and every scientific dataset of animal behavior and ecological role, is a mapping of other species into the dataome.

Modern genomic and epidemiological data on pathogenic bacteria and viruses is also abundant. Once that information is encoded in the dataome it can assert itself like any other. Uploaded information on viruses and microbes can, indirectly, exert influence on other species by altering patterns of hygiene behavior or drug production. The advertisement you see extolling the virtues of washing your hands seems like it will be detrimental to pathogens, but only to those relying on your lax toiletry habits. Others, perhaps airborne, could benefit from our changing behaviors.

Extraterrestrial dataomes, further along in their evolutionary tale, could be even more the products of multiple minds, and multiple species, binding those entities together through time. A billion-year old dataome might contain the informational fossils of untold numbers of species that have come and gone, evolving into new forms or simply becoming extinct—a kind of library of the life of a world, analogous to the library we all share with our DNA and its pathway back to our last universal common ancestor. But as a dataome it could be much more: an eidetic record not just of genes and traits but of thoughts and creations, data and culture.

It is also possible that a biology-machine holobiont would eventually come to terms with the core nature of its existence. That would mean accepting that the true meaning of existence is aiding the flow of the universe toward a state of thermodynamic equilibrium, an endless Great Boredom. Such a holobiont would understand that the generation of ever more information instantiated in ever finer grades of the material universe is a kind of cosmic duty in aid of energy dissipation. The ultimate way for such an entity to be at peace and pay homage to the laws that created them is to speed the way to being at one with those same laws.

But is there actually an endpoint? What could a very, very old dataome actually look like? One possibility is that it would have expanded beyond the physical confines of any planet of origin, both in order to utilize and dissipate more energy and to accommodate the kind of exponential growth we see in our own computing technology today. This scenario has long been a subject of speculative play for scientists and futurists.

The most famous example is arguably a paper written by the extraordinary physicist Freeman Dyson in 1960. In this single-page gem, Dyson ran the numbers on what it would take to fully accommodate a civilization like ours that grows at 1 percent per year. It doesn't sound very challenging, but compound interest is very deceptive. That rate of growth ends up as a trillionfold enlargement after just 3,000 years.

Dyson's solution for this bursting-at-the-seams civilization was a fantastically over-the-top proposal, which nonetheless looks quite innocent written on a page. If, he said, the entire energy output of the Sun (not just the bit that Earth intercepts) could be captured for around 800 years, that would be enough to somehow disassemble the planet Jupiter and use its matter to build a two- to three-meter-thick shell surrounding the Sun at the distance of Earth's orbit. Lo and behold, on the inside of this sphere you'd have lots of real estate (a healthy 550 million times the surface area of Earth) and lots and lots of solar power.

The other attribute of this modest structure, now known as a Dyson Sphere, is that it too would participate in the dissipation of energy and the growth of entropy as I've described. What comes in from the Sun must eventually go out again on the cooler exterior of the sphere. But just as energy is processed by life and Earth into thermal, infrared radiation with higher entropy, the sphere would do the same on a much grander scale, sitting there like a giant warm ball in the cosmos. The same imperatives of dissipation and entropy growth could naturally drive life toward this kind of barely imaginable engineering.

Speculation of this ilk has spawned many offshoots. The physicist Fred Adams at the University of Michigan has shown that it could be mere decades before the mass of all data storage on Earth actually exceeds the total mass of the biosphere. He's also gone on to posit how a species might "mine" older stars for graphite and silicon in order to build the computational equivalent of a Dyson Sphere—what he's dubbed a "black cloud computer."

Perhaps, though, a very old dataome, blended with whatever biology its evolutionary path opted for, wouldn't be localized at all. The example I gave earlier of the migrating birds and arctic terns could provide a template, albeit scaled up somewhat. If a dataome, or metalworld, extends beyond its planet of origin it could do so in the same way that John von Neumann envisioned the workings of automata, or self-replicating machines, spreading into the cosmos like an unstoppable cloud. And just as for the artic terns, the optimal strategy for balancing energy supply and demands could involve a long and convoluted journey, spending time here for energy, there for resources, and so on. All ultimately in unwitting (or passionate) service of the dissipation of energy and the furtherment of entropy.

There are even more fantastical possibilities if we give ourselves freedom to extrapolate what alien minds could be capable of. The Landauer limit (as well as the speed of light) implies that there has to be an ultimate barrier to the efficiency of computation and the rate of utilization of data anywhere in the universe. But a species smart enough to approach that limit with its technology would also be smart enough to see a way around it.

Here's one way that could work: The physics of Einstein's General Relativity tells us that both space and time are flexible properties. A clock running at Earth's surface runs slower than a clock sitting in empty space—because the mass of Earth warps space and time. The most extreme concentrations of mass in the cosmos, those enigmatic objects we call black holes, also represent the most extreme warping of space and time. Anything approaching the point of no return, or event horizon, surrounding a black hole will have time slow down to a near standstill with respect to the passage of time out in the rest of space. Consequently, if you parked yourself close to an event horizon you would see time pass in the surrounding universe at a hugely accelerated rate.

Imagine, therefore, that you are part of a highly technological species that has maxed out all of the possibilities for improving computation and information processing. Perhaps your species has covered planets and star systems with its equivalents to our Mongolian data centers. But in the end the physics of Landauer's limit and the finite speed of light prevent that number-crunching from further development.

The solution is to take your entire civilization and move it to just outside the event horizon of any convenient black holes (the giant ones at the center of most galaxies, for instance). As you do this, your data centers and metalworlds—left behind elsewhere in the cosmos—will speed up from your perspective; because your clock is running slower and slower, theirs will run faster and faster. Depending on how close you can place yourselves to the event horizon, you can make those external computing resources work arbitrarily fast for you. Because you remain outside the event horizon, you can still communicate with those resources, receiving and sending data, or sending probes back and forth. The main challenge is how to handle the data rates and the changing frequency of electromagnetic radiation as it falls into your gravity well or climbs out again.

The secondary challenge, a piddling thing, is that the universe itself will continue to evolve, potentially thwarting even the most grandiose schemes, but I'll talk about that in a moment.

What is critical to recognize about this kind of awesomely speculative future for us or for any other thinking species is that there might be no clear decision making at any stage. No one will wake up one morning—or no machine will click into action—and think, "We must go off-world and spread across the cosmos!" Instead, it would all just happen as a series of evolutionary steps. Those steps, as we've seen, might be pretty obscure to anyone living through them. But all will be driven by the same logic of self-propagation, of "what we see is what survives," that sits at the root of life on Earth and our present dataome. That and thermodynamics.

It is possible that we'll manage to catch up with evidence that some of these outcomes and technologies—from modest to extraordinary—have happened elsewhere in the cosmos, before it all happens to us. But on an even grander scale, there are questions about the ultimate future of all dataomes in the universe, ours included. From more than a century of astronomical observation and theoretical cogitation, we've realized that the universe has a finite age, of around 13.8 billion years, and it is expanding—the fabric of space is growing so that distant galaxies appear to be racing away from us. Furthermore, it seems that the rate of expansion is itself accelerating.

We've also learned that the kind of universe that we inhabit has a finite amount of astrophysical action it can muster. Stars will not keep being born forever. We're already well past the peak of stellar birth, which occurred in an extended flurry centered around 10 billion years ago when these objects were being popped out of condensing intergalactic gas across most galaxies. Eventually, as far as we can tell, the ever-expanding cosmos will evolve toward a condition of thermal uniformity and dullness (that Great Boredom). This hypothesis is motivated by our old friend the second law of thermodynamics, and the continual tendency for the dissipation of energy to convert useful energy into disordered, high-entropy, thermal energy. The universe may wind up as a chilly bath everywhere, a perfect, and perfectly useless, state of equilibrium. We call this the Heat Death of the cosmos.

At that future point all stars will have long ceased shining with the power of nuclear fusion. The very last of their kind, the lowest-mass and most slowly evolving specimens, will have possibly lasted until some 100 trillion years from today. Eventually the remnant parts of even those last stars and any of their planets will glow away whatever energy they have in excess of their cosmic surroundings. Beyond this point our projections are barely more than conjectures. There is a chance that the nuclear constituents of matter, such as protons, are unstable on timescales of around a trillion trillion trillion years. In that case the fabric of atoms and molecules as we experience them will also begin to dissipate into photons and particles like electrons. And if that doesn't happen, one alternative is that over time (a mere $10^{1500}$ years from now) the matter in stars and planets could actually transmute via quantum tunneling into ordinary iron and then potentially collapse into black holes (which will then very slowly evaporate to more photons thanks to the phenomenon called Hawking Radiation).

What then of information, of all the dataomes that might have existed across the universe? That informational residue would also eventually dissipate and evaporate, with no useful energy left to do the work of utilizing the data. But in the much less extreme reaches of the cosmic future we might imagine remnants of even the present human dataome. We have, after all, already sent out robotic probes like the Pioneer and Voyager missions that encode a modest amount of our information, deliberately imprinted as etchings in their "golden records" and inadvertently in their functional forms. These machines could persist for a very long time, tens of thousands or even tens of millions of years, before the erosive properties of radiation and interstellar gas and dust grind them down. In this sense we already know that humans as we are today will be outlived by this piece of our dataome, because biological evolution is likely to wreak drastic changes to us well before those interstellar probes disappear.

But if the trajectory we see on Earth—of the blending of algorithmic, informational representatives, be they genes or machines—is replicated across the cosmos, the typical longevity of meaningful, functional information in the universe could be very significant. Indeed, just as stars were forming in a fantastic frenzy billions of years ago, there might be an analogous history and future for information. A cosmic "information formation" history, if you will. An intriguing question is whether this history has a peak and where that peak might lay. Have humans come before, after, or during it?

There's one other twist to these dreams of a cosmic future. It's a story that goes back to the Austrian physicist Ludwig Boltzmann, who played a pivotal role in the development of statistical mechanics back in the latter part of the nineteenth century. The essence of it is that if you try to apply the ideas of statistical mechanics—the notions of microstates and macrostates, and of probabilistic fluctuations in those states—to something like the universe as a whole (and why not?), you can run into some disconcerting propositions.

Perhaps the most unsettling proposition stems from asking whether the conditions of our entire universe are themselves a statistical fluctuation from some underlying state of thermal equilibrium. What if our reality is just an unusual blip in a statistically boring and much larger universe that's already in a heat death state? And that blip is a fluctuation downward to a *lower* entropy state, coinciding with our Big Bang, and what we see of the cosmos is laboriously making its way back up from that?

Now, lots of physicists and cosmologists over the years have objected to this picture, and one of the most entertaining ways they've done this is via the concept of a "Boltzmann Brain." In basic terms, the more extreme that fluctuations are from some equilibrium, the rarer they are. So, a fluctuation to make a whole universe like ours has to be extremely rare. Yet that implies that a fluctuation that merely causes a single functioning brain (with memories and perceptions like ours of being in the universe) to spontaneously appear by, say, the chance proximity of atoms, must be much *less* rare.

Given the present age of the universe as it appears to us, and the extraordinarily long future that it may have, it follows that it would be more likely that you or I are actually Boltzmann brains that just think they know what's going on. Indeed, it's even more likely, in this picture, that most brains are disembodied and adrift in a greater universe (an infinite one in the original proposal). Those poor brains are isolated, cold, and presumably short-lived. The whole idea is similar to the old trope of a long-lived monkey randomly hitting keys on a typewriter. Given an infinite time the monkey would produce the works of Shakespeare, along with everything else.

The seemingly absurd nature of this is what makes Boltzmann brains part of the argument *against* the idea that our universe is a low-entropy blip in something much vaster. But it seems to me that there is a loophole for Boltzmann brains (not universes per se), because of the potentially enormous span of future awaiting our reality—most of it in thermal equilibrium, with occasional rare fluctuations. That implies that even in heat death, the universe could spontaneously produce brains, dataomes, and information—including exact replications of the past, present, and future human dataome. Not that they'd last long.

To put this another way: the last meaningful things that occur in the universe could be the randomly reconstructed classic informational hits of all thinking species that ever talked, sang, wrote, and generated a dataome, as well as species that never, in actuality, existed at all. If that seems pretty far-out and crazy, there's something even more mind-bending to come. All of these proposals rest on the assumption that everything that has happened, is happening, and will ever happen is describable with rules governed by the nature of randomness. But there could be utterly unexpected and unpredictable things going on in our universe that are *not even* random.

## Is the Universe Open-Ended?

One of my favorite, albeit heavily paraphrased, quotes from Albert Einstein is his assertion that the most incomprehensible thing about the universe is that it is comprehensible. (What he actually said, in his 1936 work "Physics and Reality," is more long-winded, and includes a digression into Immanuel Kant and the meaning of "comprehensibility," but he does write ". . . the eternal mystery of the world is its comprehensibility.") In truth, this statement holds back a little. The greater mystery is that the universe is actually capable of self-comprehension.

From a time nearly 14 billion years ago when all matter and energy existed in an exquisitely uniform and boring state, the cosmos has evolved to contain complex structures that—in at least one tiny spot in our solar system—have gained mysterious things like agency and consciousness that compel them to try to decode reality. In doing so they (meaning we) also produce interpreted versions of reality that they place in a dataome. Every equation of physics, or every computer simulation of how planets, stars, and galaxies orbit and evolve, is a bizarre imprint of an interpretation of the universe by the universe, built into the universe by the rearrangement of its atoms into brains, books, and hard drives. The question is whether or not this was all really inevitable. Did we ever have a choice in the way things have gone, and does any self-aware entity in the universe have a choice?

In a wonderfully lively, and extraordinarily ideas-dense, near-seventy-page-long essay titled "The Ghost in the Quantum Turing Machine," the theoretical computer scientist Scott Aaronson goes deep in search of arguments for and against such free will. It's such fun that I want to spend some time with it here. He points out that many of us conflate the idea of random unpredictability with free will. For example, I can feel like I'm exerting free will if I, well, I don't know, spontaneously write the word "sponge" here. It certainly seems entirely random.

That, Aaronson argues, is probably not right because what we call randomness actually follows well-defined statistical rules of probability, and in that sense is never "free." Its unpredictability is predictable. By contrast, there is a class of unpredictable phenomena that can't be measured by random probabilities; they have a different form of unpredictability. This is described by a property called *Knightian uncertainty* after one Frank Knight, an economist working on these ideas in the 1920s. In modern vernacular this is very much like the "black swan event" idea popularized skillfully by the writer and mathematical thinker Nassim Taleb. A black swan event is extremely rare, impacts the world greatly, and has explanations invented for it after the fact. But if that event or behavior can't ever be objectively quantified by probabilities, it's likely in the category of Knightian uncertainty.

Here's an example based on Aaronson's explanation: Imagine that a computer program generates random numbers as part of its operation. Perhaps it's picking random color mixtures for its screen saver. But if it picks the number 669988, there is a bug in its code that will cause it to crash. The original programmer knew this, but—since 669988 is merely 1 choice out of 1 million possibilities for this six-digit number—decided those were acceptable odds.

However, what if the code instead asks a human to provide a random six-digit number? The programmer cannot possibly know how likely it is for 669988 to be input. It could be a person's lucky number, or there could be some weird human predisposition to these digits. Instead of being predictably unpredictable it is simply unpredictable, and cannot be described by straightforward mathematical probabilities. Instead it reflects the free will of a human being.

But if you are a physicist (or a proper philosopher), you might pick a fight with this. That's because, you'd say, what a human does at any moment is ultimately a consequence of a very long, very complex chain of events. Each of those events can be broken down to individual interactions and occurrences of atoms and electrons, photons, and laws that—even if probabilistic—do still describe all options at all times; they're all predictably unpredictable. And that includes things like quantum uncertainty. Surely we can always explain a human action, or anything else, by simply going far enough down this chain of random things. In this case, there is no genuine free will; no real Knightian uncertainty in the base pieces of reality.

Aaronson argues that if the very earliest (quantum) state of the universe has Knightian uncertainty, then things are more interesting. The precise state of the new universe need not be determined by the statistical rules of randomness. It could be just as weirdly unpredictable as the previous example of someone perversely guessing the code-crashing number. In this case, the information that describes that state—and subsequently all states that the universe will take on, including all of its atoms, us, and any aliens—can be considered (in Aaronson's terminology) as being made of "freebits." And freebits are kind of like the last word in cosmic choice.

These freebits also have to be quantum in nature. That means they are also "qubits"—the version of plain old 1 and 0 bits that applies to objects and systems exhibiting quantum behavior. They are fuzzy, undetermined things until called upon and snapped into focus. That's a complication that I'm going to avoid really dealing with, because it will really make our heads hurt. Luckily, to get a sense of where freebits lead us doesn't require knowing all of those details.

The story to pay attention to is simply that these freebits could stick around throughout the history of the universe. Or, to turn this the other way: suppose you want to track back the chain of events that led to a specific incident—something interesting in a physics experiment, or a chicken crossing the road. For some incidents there will be a chain that goes *all the way back* to the original freebits. And because those freebits obey Knightian uncertainty, there is no ultimate answer for why you saw what you saw, no neat and tidy final, probabilistic solution. It will never, ever be known why the chicken crossed the road.

That could, perhaps, also apply to structures like the human brain and its thoughts. If we could disentangle the untold quadrillions of molecular and atomic interactions and chained events in a brain, and the ever-so-subtle nudges of quantum uncertainty here and there, we might find that it all leads back to the original freebits, thereby restoring some kind of free will to ourselves. I'm not suggesting any sort of daft mystical quantum-brain connection; this is all just physics (well, all physics-at-the-boundary-with-philosophy). But it could well be that your spontaneous decision to place an unsuspecting chicken at the roadside is truly Knightian, with a lineage going all the way back to the Big Bang.

Yet this also implies that there are only so many ways that anything can happen in the cosmos, only so many ways that history could unfurl. It's rather like taking a cross-country road trip from one continental coastline to another: you only have so many places you can start from, and each will influence where you end up. Is the universe truly open-ended in its capacity to generate informational novelty? Perhaps not entirely.

You might, if you've survived reading this far, be wondering how many freebits we're talking about. After all, the knowable universe is big but decidedly finite. We can only ever observe the realm of the cosmos from which light has had time to reach us since the Big Bang some 13.8 billion years ago. This is tricky, but we can actually estimate the maximum number of any kind of bits (not just freebits) in the observable universe as approximately ten to the power of 122 (or $10^{122}$). The implication is that this is the limit of the number of interesting things that can *ever happen* in the universe. No do-overs, no extras, this is it.

But this also means that freebits, and bits, are getting "used up" over time. Indeed, they must be for events to occur. And this brings us full circle back to the classical physics ideas of the laws of thermodynamics and entropy, and the Landauer limit on energy needed to erase bits. Storing and accessing information means using energy. But if you use energy you have to maintain or increase the entropy of the cosmos (generally speaking). If there are a finite number of bits in all of reality, even if a huge number like $10^{122}$, then eventually the universe runs out of ways to change its entropy, and its bits.

At this point the story reconnects to that far, far, far cosmic future in which everything is in thermal equilibrium: Space is at the same temperature, everywhere. There are no hot and cold spots, no ways for energy to flow from warm things to chilly things. No more bits to flip and the universe ends up as a tepid bath, full of nothing but regrets. (Although regrets imply information, and there would be no way to access that at this late stage.)

Is any of this a valid description of the world that has been and is to come? We don't really know, although our best bet is that the ever-expanding universe is indeed heading to eventual boredom in thermal uniformity. Concepts like freebits are, for now, merely intriguing proposals about what makes reality tick under the surface.

The essential point to all of this is that, once again, information shows itself to be more than one might expect. It isn't just a way to probe the fundamentals of nature; it may be part of the fundamentals. Consequently, the fact that the human dataome is becoming increasingly entwined with the fabric of the universe—as pieces of manipulated matter and energy—means that we (as living things) are fully committed to the universal drive toward that future ocean of unchanging, equilibrated space-time. It is as if we popped out of the vacuum as a temporary fluctuation of energy, and we've been clawing our way back ever since.

## The End, Part One

It might seem that in drawing together the many threads of this story, I'm landing in a position of resignation to the inevitability that our species cannot really help itself because of the inexorable and insurmountable drive of fundamental physical laws in this universe.

Or, to put this a different way: the hours you spend staring at tiny screens and social media, our species' ever-dwindling attention space and ability to discriminate truth from post-factual madness, are all to be expected. The psychologically fine-tuned nuances of screen swipes, fifteen-second video highs, and the resulting absurd wealth inequalities are, seen from a higher perspective, all in service of lubricating the movement and generation of information, of datastreams happily dissipating energy to the cosmos. It is an ultimate nihilism, where the meaninglessness of life has meaning.

That is, to say the least, a little depressing. Does the dataome, which has helped lift us from a state of modest sophistication to a state of sublime brilliance—into a species capable of decoding the mechanisms that evolve a universe of random nothingness into stuff with the complexity of differential calculus, or the glorious works of Mozart, da Vinci, Mary Cassatt, Frida Kahlo, and Banksy—really just drive us to a future of more entropy and planetary disaster?

The human end-story I have here is not entirely scientific, but it is perhaps realistic. Our species has emerged in a window of opportunity, after an age of giant, terrible lizards on a planet dodging between moderate ice ages and a sticky, humid greenhouse. The ongoing experiment of biological evolution has, this time, landed on a trick in the dataome that is both new and perhaps entirely inevitable. Whether or not that carries us further than it has already is unclear, and depends a great deal on how we balance the demands of the dataome with our biological needs for a stable and nurturing planetary environment.

A common rebuttal to the evidence for our increasingly overwhelming energetic burden in digital data and computation—surveyed in earlier chapters—is that we are also getting more and more efficient. Von Neumann's EDVAC in the 1940s ran a few thousand calculations a second, but occupied a space of 30 by 14 feet, and at its peak consumed 56,000 watts of power. By comparison, a 2020 iPhone central processor (with some 8.5 billion transistors) can churn through 155 billion floating-point operations a second and consumes around 1 watt of power running at that level.

But, as I've discussed, we also grow our needs and expectations for data in a way that seems to perpetually overwhelm any of these remarkable efficiency gains. We live in a world with certain giant multinational corporations whose central business models depend acutely on how much data flows, and how engaged humans are with that torrent. If you want an example of the emergent "song" produced by a holobiont, then large tech companies might fit the bill. They compete intensely with each other, are selected for by their profit and anticipated profit, and their very existence is a result of the combined and competing interests of the dataome and humans.

Faced with this situation, how can we both preserve and support our dataome and preserve and support the planetary environment that is so essential for our lives? The most obvious answer is that we need to treat information as a natural resource, much like I've argued it to really be. And as with any other natural resource, from fresh water to crude oil, or oxygen and sunlight, we cannot extract it, refine it, or utilize it without cost or repercussions. Information really isn't "free," nor has it ever been so.

However, managing information as a resource feels very unpleasant and counterintuitive. After all, we'd not want to say to a budding Shakespeare that they should slow down their writing, or do less to promote a readership. To avoid that kind of tension it seems that we have to concentrate on how we choose to *implement* information, the ways that information is instantiated in the physical world outside of our bodies. More broadly, there may be arguments to resolve over whether the dataome is best treated as a part of a holobiont (akin to our microbiome), or whether it is better to treat the dataome's health and growth as we would manage any staple crop, from soybeans to rice and wheat. They too are resources, but they rely on other fundamental resources, from water and soil to sunlight.

In either approach—"biological" or "agricultural"—identifying the markers of dataome health or behavior is going to be challenging. While it's pretty clear that the meaningfulness of information within the dataome is one critical test of success (especially as it pertains to survival), it's far less clear how to practically evaluate that, or how to incorporate something like a dataome agriculture into our present economic principles and structures. Some nations have implemented an indirect carbon tax as a price instrument in an effort to curtail the use of hydrocarbon fuels. It might be possible to use that same framework to add a dataome tax, assigning a price to the "emission" of information, based on the conventional resource burden—from raw materials to manufacturing to pumping electrons through circuits, with higher tax rates applied to data of lower utility or integrity.

It is also possible that we simply don't understand enough yet about the dataome, and our intricate relationship to it, to be able to make these decisions. Much as with our microbiome, we're only just beginning to recognize the role the dataome plays in our lives and our evolutionary pathways. Our species is not fully defined by its biological forms, and we may be already sharing Earth with another form of life. For these reasons alone we may no longer be simply asking how best to ensure the survival of *Homo sapiens*, but rather the survival of a symbiotic system. That system appears to be able to evolve rapidly, over years and decades, but also on much longer timescales. It would be wise to be thinking about our continued coexistence at least a thousand years out from today.

One thing is clear, though: our chances can only improve if we question the world, build our knowledge, aim for wisdom, and do the best we can for the generations that we hope come after us. That means celebrating and supporting all of our minds, those special constructions of neurons that represent the most complex systems we presently know of. That also means weeping at the loss of malnourished children in war-torn countries and objecting to the pathetic shortsightedness of too many of our leaders and advisers. Humanity does not belong to those failures, it belongs to our blended selves, biology and machine, data and mind. It's time to embrace that future.

## The End, Part Two

I want to conclude this book on a much more down-to-earth level. We've followed a long trail from the relatively straightforward observation of Shakespeare's informational legacy and the energetic burden of all our data to an extremely high conceit: the idea that ideas are a fundamental currency of the universe, and that we humans are an expressive expression of that fact.

But what about you and the part you play in all of this? Think about your life and all the ways in which you participate in the dataome. There were, for example, your very first words, likely learned from parents and family as your juvenile neurons were exposed to an endless barrage of sensory inputs from the world. Those symbol clusters and language structures had already propagated for centuries or longer as evolving forms, held aloft by human minds.

Along with your first words came the categorization of shapes, colors, smells, sounds, textures—many from objects that exist as pieces of a human dataome: wooden blocks, colorful books, and toy animals. All these items are not just manifested as physical things but also held as data in manufacturing blueprints, printers' files, and designers' software. And now, whatever path you are taking through life, like it or not, you are an ongoing participant in the propagation of the dataome.

Take a moment to examine your day so far. At this very moment you are interacting with the dataome by reading these words. Next you may read something else: an email, a text, a traffic sign, a bus timetable, the app for your car service, the ingredients on the packet of noodles you're about to cook. You've probably already written something today, or will. Or maybe you've left a voice message on someone's phone or passed along a meme-like thought to a friend. Perhaps you bought a pastry this morning with a credit card or electronic payment; both have inserted new data into the dataome. You are, from the perspective of the dataome, both an essential part of the machinery of information and a filthy leaky creature, splashing around in the ocean of letters, words, and bits. You simply cannot help but consume and shed data.

Over time humans, and human systems of commerce and society, have learned the value and power of that leakiness. Today, almost anyone who moves through the world is, to be blunt, being exploited for their data interaction and generation. One of the clearest articulations of this fact comes from the visionary technology pioneer Jaron Lanier. It's hard to categorize Lanier, but child prodigy, artist, scientist, and dreadlocked deep thinker all seem to fit. Lanier is also perhaps best known as the person generally considered to have coined the term "virtual reality," in the late 1980s. One of his books, *Who Owns the Future?,* published in 2013, articulates his ideas on the current situation for humans—a world in which we are all being endlessly mined for our personal data-flesh (my words). We are consistently exploited for our generation of, and interaction with, information. He proposes that a far more equitable civilization would come about if information was always traceable to its origins and if we received micropayments for use of our data. One inevitable result in such a world would be a dramatic redistribution of wealth, a smoothing of the divide from rich to poor.

I find Lanier's ideas quite compelling, and morally commendable. But through the lens of the dataome I also wonder if it is actually possible to accomplish such a change, because of the fundamental nature of the forces at play. I've tried to show how the dataome may be as much the driver of human behavior as we are of its behavior. In many instances, from memes to Mongolian data centers, we are being steered by the dataome's requirements more than ours (including the need for a stable planetary environment). Even if you think the dataome is simply an extended phenotype, it's a doozy of a one.

Natural selection can steer systems to efficiency, reducing the friction of actions or reactions. The fact that we've skidded into a world where there is so little friction between us and the flow of information being lifted from us seems like a consequence of that kind of selection. It's not clear that even the most ingenious and robust methods, like the blockchain, for linking us to our valuable data assets can substantially regulate the flow.

Our present situation hasn't just appeared. If we look into the past we can see endless precursors of the current issues with personal data. Any rulers or leaders of a society benefitted greatly by gathering information on the behavior and status of their people. One of the best examples was the Domesday Book (the Middle English spelling for Doomsday Book, as we tend to pronounce it still today), completed in 1086. At its core this was a survey of the lands that William the Conqueror (also known, coincidentally, as William the Bastard) set in motion almost two decades after the Norman conquest of England in 1066. His motivation had much to do with the threat of an impending invasion from Denmark, and a pressing need to raise funds and assess his resources.

Because of this, the survey focused on the tangible assets of the boroughs and manors of England: who owned which bit of land and what land was arable, what land was forested, and how much communities owed in taxes, rents, and military service. But the side effect was the unleashing of unprecedented information of a quite personal nature. That information altered the courses of action of rulers, landowners, and barons, of communities, and of individual lives. All long before our modern world.

In the end we are who we are, as a species and as individuals, because of the dataome. All of our interactions and contributions to the external data of humanity are, when summed, unique for each of us. You would not, could not, be you without those first words, first books, songs, or games that insinuated their way into your neurons.

But most of us do not examine our lives this way. If that were to change we might better understand why we do the things we do and why we make the choices that seem to be increasingly hazardous for the conditions that produced us in the first place. This is really not a great burden, to see ourselves like this. Far from it, I think that reconceptualizing our existence through the lens of the dataome is extraordinarily freeing, and hopeful. Instead of fearing a future blended with machines and information, we can face it with a measure of confidence, and step through yet another threshold in the four-billion-year story of life on one small planet in a vast and magnificent universe.

There is a tide in the affairs of men.
Which taken at the flood, leads on to fortune;
Omitted, all the voyage of their life
Is bound in shallows and in miseries.
On such a full sea are we now afloat,
And we must take the current when it serves,
Or lose our ventures.

(Brutus, in *Julius Caesar*, Act 4, scene 3, by William Shakespeare)

# Acknowledgments

It feels a little odd to write lengthy acknowledgments for a book that, at its core, is all about the universal impetus for generating more information and more complexity in the world. Here I go again, I think, adding another spoonful to the cascading river that is human data, nudging along our relationship with an informational, algorithmic, *process*-fueling partner-in-crime on an otherwise innocent planet.

But a major aspect of the story in this book is how we're all in this together, whether we want to be or not. In that context it makes perfect sense to expend a few more bits on thanking the many people (and emergent phenomena) who've helped me add to the growth of cosmic entropy and, dare I say it, to the increasing energetic burden of information in this small piece of the universe.

This book, much like the phenomena it describes, unfolded in a way that was surprisingly open-ended, with novelty around every corner. The central topics of information, entropy, and evolution, are so intrinsically vast and complicated that there are surely an enormous number of possible pathways to take in telling their story. In some instances, for some branches of the tale, I've followed my own preformed interest in specific scientific (perhaps even philosophical) questions and their implications. In other places I've allowed myself to scamper after interesting signposts when they've appeared and have been extremely lucky to benefit from the brains and experience of many others. I've eavesdropped on innumerable conversations and have been tolerated graciously as I poked my nose in with my endless questions.

A starting place for specific acknowledgments is to express my thanks to *Nautilus* magazine and to Michael Segal, who went along with me when I asked if they'd consider publishing a piece I'd called "The Selfish Dataome" in the late fall of 2018. I felt my request was urgent because I was excited by the insight that I was convinced I'd had (the idea of a dataome), and I wanted to claim the intellectual space. Naturally, by the time Michael was done with me, and certainly by the time I got to writing this book I was a humbler, meeker, more sober person, having seen just how rich a topic this really was.

My fabulous agent, Deirdre Mullane, of Mullane Literary Associates, has been a patient and unflinching sounding board for corralling my wild enthusiasm into something resembling a plan, and she deserves special praise. I am also indebted to Courtney Young and the team at Riverhead for taking a chance on turning my crazy idea into a real book. Courtney's extraordinarily incisive and insightful editorial work has been essential, as has her consistent enthusiasm about the whole project, giving me the confidence to plow on. All in the midst of a global pandemic no less. And Annie Gottlieb, copy editor extraordinaire, has applied her sharp eyes to yet another of my books, smoothing rough edges and holding me rightfully accountable to the facts.

For science thank-yous where do I begin? A lucky case of six-degrees-of-separation (thank you, Dave Spiegel) led to a dinner sometime in 2015 with Piet Hut that produced friendship, adventures in Tokyo, Kyoto, Princeton, and New York City, and started a wild intellectual journey that continues to this day. Through Piet's wisdom and also the Earth-Life Science Institute at the Tokyo Institute of Technology I've been able to learn about research far beyond my comfort zone, from the origins of life to the origins of consciousness (insights obtained even without beer and sake). I'd like to thank Eric Smith, Nathaniel Virgo, Nicholas Guttenberg, and in particular Olaf Witkowski, as well as many others whose paths crossed with mine in Japan: Mary Voytek, John Hernlund, Christine Hernlund, Ryota Kanai, Stuart Bartlett, Elizabeth Tasker, Sara Walker, Norman Packard, George Helffrich, and Lee Cronin; thanks for hanging out. Across other parts of the planet there have been innumerable conversations and exchanges that have, at least in fragments, found their way into the pages of this book. Among those, thanks are due to Adam Frank, Marcelo Gleiser, George Musser, Diana Reiss, Lee Billings, Bryan Johnson, Lee Billings, Adam Black, and Martin Rees.

One adventure leads to the next, and the wild-eyed, invigorating experiment to create an entirely new kind of scientific institute that was YHouse from 2016 to 2019 also generated a lot of the mental landscape out of which the idea of the dataome arose. Inspirations were abundant from so many, including Eiko Ikegami, Erik Hoel, Ed Turner, Stephen Burlingham, the irrepressible Sean Sakamoto (check out his movies and writing), as well as valued friendships with many others, including Ayako Fukui, Barnaby Marsh, and Andrew Fleming. A public talk I gave for YHouse's "Consciousness Club" in 2018 also prompted me to start thinking about what characteristic really defines a technological species, and it was their externally held information that first came to mind. I'd like to also especially thank David Krakauer and his colleagues at the Santa Fe Institute for having me visit a couple of times, and for inviting me to participate in the fantastically bold Interplanetary Festival in downtown Santa Fe.

A special nod is owed to Dmitri Gunn who is the Executive Director of TEDxCambridge. Dmitri's generosity of time through many video calls gave me an opportunity to stress test ideas and to be challenged to explain myself better, as well as learn from his wealth of insights.

At Columbia University I've benefited from the willingness of many colleagues to put up with an ongoing journey that has me bouncing from scientific field to scientific field. That support—academic and personal—has been supremely important, and I'd particularly like to thank David Helfand, Frits Paerels, Peter DeMenocal, Amber Miller, and Mike Purdy for finding ways to keep me around. I also have to thank the remarkable Ivana Hughes and the Frontiers of Science undergraduate course at Columbia, which I've been lucky enough to help teach over the years. There are elements in this book inspired by exposure to the material from that course, and through interactions with the cohort of fantastic young scientists I've taught alongside. A special thank you also goes to Hod Lipson, who appears in the book and is an unfailingly generous and affable neighbor and colleague, even if he is sometimes a little cruel to his robots.

There's a deep and inevitable debt owed in a book like this to the hundreds of articles, papers, and other books that I've read, consulted, skimmed, and enjoyed or hated. Some of these show up in the endnotes, but certainly not all or else this would be a volume of endnotes with a small book at the beginning. I have lifted ideas, snippets, and hints from many places and many times. There are fields of inquiry—especially in information theory, computation, and physics—where extraordinary work is being done on furthering our understanding of the deepest qualities of matter, complexity, and emergence. But it's esoteric, heady stuff, and so I've had to make hard decisions about what to include and what to simply pass over for the sake of writing a text that is more accessible, and a narrative that is more enjoyable. It's always tough doing this; as a scientist it's difficult not to imagine your colleagues looking over your shoulder and tut-tutting because of what you've left out or oversimplified. But letting those ghosts go is important, because the point is to tell a story to a reader who does not already know it.

Away from academia and other pursuits I am truly grateful for those I now think of as my Catskills cohort, a troupe of neighbors and dear friends who discretely observed and unknowingly provided emotional support for much of my writing process from across the garden and around the corner: Abbie, Lewis, Ellen, Mal, Randy, Joe, Lorraine,

Christine, Sean, Ron, and Pat, thanks for making the mountains welcoming and happy. Finally, as always, I cannot express adequate thanks to my family: Bonnie, Laila, Amelia, and Marina, you have once again put up with my distracted, sometimes anxious self as words flew and bits flipped. This one's for you.

# Notes

## 1. OUR ETERNAL DATA

4 **information is still here, though:** The information humans generate has occupied people's minds before, from wonderment at the persistence of our stories and records to consternation at the ferociously accelerating complexity of our information-rich world. The great writer and historian of science James Gleick wrote a masterly and wide-ranging treatment of this in *The Information: A History, a Theory, a Flood* (New York: Pantheon Books, 2011). I tread some of the same ground, but with a different slant and interpretation, and, I think, a very different goal.

4 **neologism "meme":** Dawkins coined "meme" by shortening the Ancient Greek "mimeme," meaning "imitated thing."

5 **"extended mind":** Clark and Chalmers introduced the concept in an influential 1998 essay in the peer-reviewed journal of philosophy *Analysis*: Andy Clark and David Chalmers, "The Extended Mind," *Analysis* 58, no. 1 (January 1998): 7–19, www.jstor.org/stable/3328150.

6 **call it the "dataome,":** I first used this term in an essay written for the magazine *Nautilus* in 2018: Caleb Scharf, "The Selfish Dataome," *Nautilus* 65 (October 2018), http://nautil.us/issue/65/in-plain-sight/the-selfish-dataome.

7 **Richard Dawkins, back in 1982:** Dawkins wrote a much more technical follow-up and response to *The Selfish Gene* in another book. For many scientists it is perhaps the better volume: Richard Dawkins, *The Extended Phenotype: The Long Reach of the Gene* (Oxford, UK: Oxford University Press, 1982).

8 **of the nematodes:** Crazy stuff, but true. There's some terrific research on this remarkable kind of extended phenotype: D. P. Hughes, D. J. C. Kronauer, and J. J. Boomsma, "Extended Phenotype: Nematodes Turn Ants into Bird-Dispersed Fruits," *Current Biology* 18, no. 7 (April 2008): R294–95, doi:10.1016/j.cub.2008.02.001.

8 **"niche construction":** As with all these topics, much has been written. A nice (short) review covering the extended phenotype, modern efforts to quantify the

external influence of genes, and niche construction is Philip Hunter, "Extended Phenotype Redux: How Far Can the Reach of Genes Extend in Manipulating the Environment of an Organism?," *EMBO Reports* 10, no. 3 (March 2009): 212–15, doi:10.1038/embor.2009.18.

9 **we get "it from bit":** We'll come to Wheeler's ideas again, but one of his first writings on these ideas was in proceedings from a conference in Tokyo: John Archibald Wheeler, "Information, Physics, Quantum: The Search for Links," in *Proceedings of the 3rd International Symposium: Foundations of Quantum Mechanics in the Light of New Technology, Tokyo, Japan, 1989* (Tokyo: Physics Society of Japan, 1990), 354–68. His "negative twenty questions" illustration wasn't entirely original, inasmuch as the game of twenty questions was already a favorite for information science. An example is known as Ulam's game, or the Rényi–Ulam game, after the mathematician and physicist Stanislaw Ulam and the mathematician Alfréd Rényi.

11 **the instinct to do so:** A report on one effort to study weaver bird nest-building skills: Patrick T. Walsh et al., "Individuality in Nest Building: Do Southern Masked Weaver (*Ploceus velatus*) Males Vary in Their Nest-Building Behaviour?," *Behavioural Processes* 88, no. 1 (September 2011): 1–6, doi:10.1016/j.beproc.2011.06.011.

12 **0.1 percent to 0.6 percent:** There is a wide body of work on human genetic diversity. A recent study that shows just how genetically intertwined modern humans are is: Anders Bergström et al., "Insights into Human Genetic Variation and Population History from 929 Diverse Genomes," *Science* 367, no. 6484 (March 2020): eaay5012, doi:10.1126/science.aay5012. A slightly earlier study showing the similarity across modern humans was: Jun Z. Li et al., "Worldwide Human Relationships Inferred from Genome-Wide Patterns of Variation," *Science* 319, no 5866 (February 2008): 1100–1104, doi:10.1126/science.1153717.

12 **discovered in 2012:** This was an intriguing study, well worth looking at: Rory Bowden et al., "Genomic Tools for Evolution and Conservation in the Chimpanzee: *Pan troglodytes ellioti* Is a Genetically Distinct Population," *PLOS Genetics* 8, no. 3 (March 2012): e1002504, doi:10.1371/journal.pgen.1002504.

13 **sub-Saharan Africa:** There are quite a few studies now. A comparatively early review that's a useful reference point is: Michael C. Campbell and Sarah A. Tishkoff, "African Genetic Diversity: Implications for Human Demographic History, Modern Human Origins, and Complex Disease Mapping," *Annual Review of Genomics and Human Genetics* 9 (September 2008): 403–33, doi:10.1146/annurev.genom.9.081307.164258.

13 **Tjapwurung people:** In southern Australia, sometimes spelled as Djab wurrung. See, for example: Patrick D. Nunn, "The Oldest True Stories in the World," *Sapiens Anthropology Magazine*, October 18, 2018, https://www.sapiens.org/language/oral-tradition/.

13 **Genyornis newtoni, a fowl:** There is some debate on exactly when this species disappeared, and whether humans or climate change was a deciding factor in its extinction. In this study researchers looked at burn marks on eggshells, presumably left

by humans cooking the eggs: Gifford Miller et al., "Human Predation Contributed to the Extinction of the Australian Megafaunal Bird *Genyornis newtoni* ~47 ka," *Nature Communications* 7 (January 2016): 10496, doi:10.1038/ncomms10496.

14 **Last Glacial Maximum:** A seriously chilly time, this was also a dry time globally that lasted from about 33,000 years ago until about 20,000 to 14,000 years ago. As it waned, ice sheets that had extended to low latitudes (including over the island of Manhattan) melted and receded, and there were pulses of sea level rise as polar zones shrank.

14 **multiple lines of the oral histories:** Some fantastically interesting work has been done on this; see, for example: Patrick D. Nunn and Nicholas J. Reid, "Aboriginal Memories of Inundation of the Australian Coast Dating from More than 7000 Years Ago," *Australian Geographer* 47, no. 1 (2016): 11–47, doi:10.1080/00049182.2015.1077539.

14 **Sumerian cuneiform:** The cuneiform approach to writing was remarkable for its versatility and longevity, propagating beyond just Sumerian. The cuneiform method was used also by the Akkadians, Babylonians, Elamites, Hatti, Hittites, Assyrians, and Hurrians. See, for example: "The World's Oldest Writing," *Archaeology* (May/June 2016), Archaeological Institute of America, www.archaeology.org/issues/213-1605/features/4326-cuneiform-the-world-s-oldest-writing.

16 **Sassanian Empire:** The last robust gasp of the Persian Empire before the spread of Islam, the Sassanian dynasty ruled for some four centuries, from 224 to 651 AD. It was an enormously sophisticated and complex society. See, for example: Touraj Daryaee, *Sasanian Persia: The Rise and Fall of an Empire*, International Library of Iranian Studies (New York: I. B. Tauris, 2009).

17 **George Smith:** Although it's almost certainly an oversimplification to attribute so much to his efforts, his is still a good story. See, for instance: David Damrosch, "Epic Hero," *Smithsonian Magazine*, May 2007, www.smithsonianmag.com/history/epic-hero-153362976/, excerpted from Damrosch, *The Buried Book: The Loss and Rediscovery of the Great Epic of Gilgamesh* (New York: Henry Holt, 2007).

18 **devices called "khipu":** Much is still to be understood about this remarkable form of information storage. See, for example: Charles C. Mann, "Unraveling Khipu's Secrets," *Science* 309, no. 5737 (August 2005): 1008–9, doi:10.1126/science.309.5737.1008 also Gary Urton and Carrie J. Brezine, "Khipu Accounting in Ancient Peru," *Science* 309, no. 5737 (August 2005): 1065–67, doi:10.1126/science.1113426.

20 **"tangles for their souls.":** A delightful tale in this essay on khipu and Spanish invasion (I mean colonization): John Charles, "Unreliable Confessions: Khipus in the Colonial Parish," *The Americas* 64, no. 1 (July 2007): 11–33, doi:10.1353/tam.2007.0099.

20 **Khipu Database Project:** A fantastic resource on khipu, with explanatory material (or at least particular interpretations, as is the way of research) and actual khipu data for study: http://khipukamayuq.fas.harvard.edu/.

21 **by around 800 AD:** There are many scholarly works that explore the early development of printing systems. In the case of China a nice popular-level essay lays out some of the timeline: M. Sophia Newman, "So, Gutenberg Didn't Actually Invent

the Printing Press," *Literary Hub*, June 19, 2019, https://lithub.com/so-gutenberg-didnt-actually-invent-the-printing-press/.

22 **a financial disaster:** Depending on the sources you read, you'll see Gutenberg either feted as a genius or described as a shifty, deceptive grifter. It certainly doesn't seem that all of his story is particularly well-documented or clear, with some gaps in time and uncertainty about just where he gained his skills. All of which is delightfully fascinating given the association of his name with a printing technique that has arguably done so much for the consistent, precise storage of human data.

## 2. THE BURDEN OF AN IDEA

25 **835,997 words:** Seemingly useless facts like this can be obtained from excellent resources such as OpenSourceShakespeare (www.opensourceshakespeare.org/). Here you can search the texts and extract all manner of statistics. For example, across all of Shakespeare's works there are 12,493 unique word forms that only occur once (he liked his word creation). Falstaff is the champion of speeches, with 471 in total. The word "sponge" occurs exactly 4 times in total.

25 **two to four billion:** These are very rough guesstimates. But if you take 37 plays (for instance), and 400 years of printing, you only need to print at a rate of about 22,500 copies a month to reach 4 billion total copies. Split among, say, 20 countries as the prime producers, this is a mere 1,100 or so per month per country.

26 **making physical books:** Pulling together these numbers is not too hard. Some resources include: the "Our World in Data" website from the Oxford Martin School at the University of Oxford (https://ourworldindata.org/books), which also makes use of data compiled by Eltjo Buringh and Jan Luiten Van Zanden in "Charting the 'Rise of the West': Manuscripts and Printed Books in Europe, a Long-Term Perspective from the Sixth through Eighteenth Centuries," *The Journal of Economic History* 69, no. 2 (June 2009): 409–45, www.jstor.org/stable/40263962. Modern book sales data is freely available from a variety of sources, including *Publishers Weekly*.

26 **over 4 trillion joules of energy:** Most of us aren't used to measurements of energy per se. But we might have some intuition for what something like 60 watts means in a relative sense. One watt is a unit of power, or a joule per second—the amount of energy being used or deployed in a given amount of time. I've used joules and watts because it gets increasingly contrived to express energy and power in equivalent terms (although I've done that here too, in comparing to combusted coal).

27 **expansion from Africa:** Here I am assuming the commonly quoted figure of the first major movement of *Homo sapiens* out of the African continent having occurred some 60,000 years ago. Some recent estimates push evidence of *sapiens* living outside of Africa as far back as 185,000 years. Whether that was due to a major movement or not is unclear.

27 **US paper production:** Digging around for these numbers is challenging, but a thorough and useful resource is "The State of the Paper Industry: Monitoring the Indicators of Environmental Performance," a collaborative report by the Steering Committee of the Environmental Paper Network, 2007, full text online only, https://environmentalpaper.org/wp-content/uploads/2017/08/state-of-the-paper-industry-2007-full.pdf.

28 **printing ink are produced annually:** I used some of the figures quoted by the testing and consulting firm Smithers in a commercial study on "The Future of Global Ink Markets to 2023": see www.smithers.com/Services/market-reports/Materials/the-future-of-global-ink-markets-to-2023. The figures are in broad agreement with other estimates.

28 **of high-quality coal:** Multiplying this mass by a factor of 2.8 yields a carbon dioxide output. This factor for coal combustion is available in many sources, including B. D. Hong and E. R. Slatnick, "Carbon Dioxide Emission Factors for Coal," US Energy Information Administration, *Quarterly Coal Report,* January–April 1994 (Washington, DC: DOE/EIA, August 1994), 1–8.

29 **about 2.5 quintillion bytes:** Quoting any specific figure for this is tricky, since it's both hard to estimate accurately and constantly increasing. See, for instance, this World Economic Forum site (April 2019): www.weforum.org/agenda/2019/04/how-much-data-is-generated-each-day-cf4bddf29f/. And a reminder that a quintillion is $10^{18}$, so 2.5 quintillion bytes a day is 2.5 exabytes a day—a figure that may have grown by a factor of 10 or even 100 by the time you read this.

30 **"1.7-kilogram microchip.":** Although now almost twenty years old, this remains a highly informative and relevant study. Eric D. Williams, Robert U. Ayres, and Miriam Heller, "The 1.7 Kilogram Microchip: Energy and Material Use in the Production of Semiconductor Devices," *Environmental Science and Technology* 36, no. 24 (October 2002): 5504–10, doi:10.1021/es025643o.

31 **the judgmental parlance:** I think it's accurate to see the terminology and approach to science by those in certain schools of thought being influenced by their society. In the 1800s, the Protestant ethic of virtue, merit, and hard work insinuated itself into the formative years of the field of thermodynamics.

31 **second law of thermodynamics:** If you've not studied physics you might wonder what the first law is and whether there are others. Be comforted that there are three such "laws." The first is simply the conservation of energy—it can neither be created nor destroyed in an isolated (closed) system, only turned into other forms. The third is that as a system approaches absolute zero (-273 degrees Celsius, the minimum possible temperature in the universe), its entropy must approach a constant value (which may be zero). Sometimes people invoke a "zeroth law," which is important in the mathematics of thermodynamics and is, put simply, if two things (systems) are each in thermal equilibrium with a third thing (neither is gaining or losing to that third) then they must also be in thermal equilibrium with each other.

32 **of sorting boxes:** I think this is a helpful analogy. It at least works for classical physics, where we consider things like positions and velocities that have precise values. It gets more complicated when we allow for quantum physics and uncertainties.

33 **total number of microstates:** More specifically, in this case entropy is proportional to the logarithm of the number of microstates.

33 **kind of disorder:** Physicists have long used the term disorder to describe a quality of entropy, but not everyone agrees that it's helpful because of the everyday connotations. Some prefer the idea of "dispersal" rather than disorder—suggesting the increase in microstates among which atoms or molecules (or anything else) are spread.

34 **yield no work:** This is not an obvious statement to make, and I'd like to thank Dr. Eric Smith for reassuring me that it is nonetheless valid.

36 **in the Congo:** Tantalum and the element niobium are extracted from a mineral ore, columbite-tantalite, or coltan for short. The United Nations has reported that in the 2000s the smuggling of coltan for export helped fuel the war in Congo. There are efforts to take tantalum "out of the loop" of conflict, by finding ways to control and secure its export.

37 **as far as the 1650s:** For example, see the US Energy Information Administration (EIA) at eia.gov and its Annual Energy Review www.eia.gov/totalenergy/data/annual/archive/, which includes historical data in the annual reports (with summaries back to 1635).

37 **By the year 2000:** An excellent resource is Our World in Data https://ourworldindata.org/energy, a collaboration between researchers at Oxford University's Oxford Martin Programme on Global Development (at the Oxford Martin School) and the nonprofit Global Change Data Lab.

37 **47 billion watts:** There are many sources with estimates of energy consumption by computation and data services, and the numbers keep on changing upward. See for example Nicola Jones, "How to Stop Data Centres from Gobbling Up the World's Electricity," *Nature* 561 (September 2018): 163–66, doi:10.1038/d41586-018-06610-y. Also projections by the Semiconductor Industry Association, "2015 International Technology Roadmap for Semiconductors (ITRS)," www.semiconductors.org/resources/2015-international-technology-roadmap-for-semiconductors-itrs/.

38 **located in Hohhot:** Strictly speaking the data centers are 10–15 miles outside of the town, itself an increasingly bustling capital, with a population of around three million.

38 **Inner Mongolia Information Park:** This center alone consumes about 150 megawatts of power on average.

39 **increases in efficiency:** Efficiency improvements do help, and the picture may not be as dire as feared a few years ago, but the amount of computation is growing too: Eric Masanet et al., "Recalibrating Global Data Center Energy-Use Estimates," *Science* 367, no. 6481 (February 2020): 984—86, doi:10.1126/science.aba3758.

40 **energy use per computation:** For several decades up to around the year 2000 the number of computations per unit power was doubling every 1.5 years. But after 2000 that rate of improvement slowed significantly, until seemingly picking up again around 2010. Exactly what the trend line really looks like is somewhat unclear, and it may be showing signs of bending down; see, for example. Figure 17 in: Jennifer Hasler and Bo Marr, "Finding a Roadmap to Achieve Large Neuromorphic Hardware Systems," *Frontiers in Neuroscience* 7 (September 2013): 118, doi:10.3389/fnins.2013.00118.

40 **in 1998 researchers:** Simon B. Laughlin, Rob R. de Ruyter van Steveninck, and John C. Anderson, "The Metabolic Cost of Neural Information," *Nature Neuroscience* 1 (May 1998): 36–41, doi:10.1038/236. Also, an interesting analysis is made by Biswa Sengupta and Martin B. Stemmler, "Power Consumption During Neuronal Computation," *Proceedings of the IEEE* 102, no. 5 (May 2014): 738–50, doi:10.1109/jproc.2014.2307755; US Department of Energy, Office of Science and Technical Information (OSTI), www.osti.gov/servlets/purl/1565222.

41 **neuromorphic chip design:** An area making some rapid and remarkable progress; see, for example: Mike Davies et al. "Loihi: A Neuromorphic Manycore Processor with On-Chip Learning," *IEEE Micro 38, no. 1* (January/February 2018): 82–99, doi:10.1109/MM.2018.112130359.

42 **as old as human trade:** Barter was probably the first "currency," whether in the form of exchanging physical objects or promised actions. That's fascinating in the context of the dataome because barter is inherently about information: our assessment of what an object can do for us (versus the thing we might be giving up), or our anticipation of what someone else might do for us (or not). Aristotle thought about this in his *Politics* in 350 BC.

43 **devices like ciphers:** Our friends the Mesopotamians may lay claim to an early encryption of data. Evidence exists on a clay tablet from 1500 BC, where cuneiform symbols were simply swapped for each other to obscure what was written. The secret? The recipe for a coveted pottery glaze. See, for example, a nice popular article: Kaveh Waddell, "The Long and Winding History of Encryption," *The Atlantic*, January 13, 2016, www.theatlantic.com/technology/archive/2016/01/the-long-and-winding-history-of-encryption/423726/.

43 **presents a secure model:** The paper was not published in the conventional sense, but was shared online: Satoshi Nakamoto, "Bitcoin: A Peer-to-Peer Electronic Cash System," October 2008, https://bitcoin.org/en/bitcoin-paper.

44 **The whole tale:** Goodness me, where to start? A simple internet search on Satoshi Nakamoto will set you off on the path. Where that path leads is pretty unclear, but if you want to pass the time on a rainy day, then it's all yours.

45 **Marshall Pease published:** The paper is Leslie Lamport, Robert Shostak, and Marshall Pease, "The Byzantine Generals Problem," *ACM Transactions on Programming Languages and Systems* 4, no. 3 (July 1982): 382–401, http://people.cs.uchicago.edu/~shanlu/teaching/33100_wi15/papers/byz.pdf.

47 **67 terawatt hours:** Figures like these are notoriously difficult to come up with, since they depend on a number of assumptions and extrapolations. Nonetheless, there are some brave souls attempting this. One example is a project created by Alex de Vries called Digiconomist that seeks approaches to "exposing the unintended consequences of digital trends": https://digiconomist.net/bitcoin-energy-consumption.

48 **use of personal blockchains:** I had never really thought about this as potentially important until I read this excellent article by Steven Johnson: "Beyond the Bitcoin Bubble," *The New York Times Magazine*, January 16, 2018, www.nytimes.com/2018/01/16/magazine/beyond-the-bitcoin-bubble.html.

49 **artificial intelligence, or AI:** I felt it important to be pedantic here because we all tend to fall into the popular-language use of "AI" out of convenience.

49 **2017 AlphaZero:** An amazing piece of research. David Silver et al., "Mastering the Game of Go without Human Knowledge," *Nature* 550 (October 2017): 354–59, doi:10.1038/nature24270.

49 **2019 study by Emma Strubell:** The paper is by no means the last word on the subject, but it did stir up a lot of discussion. It is in a conference proceedings: Emma Strubell, Ananya Ganesh, and Andrew McCallum, "Energy and Policy Considerations for Deep Learning in NLP," *Proceedings of the 57th Annual Meeting of the Association for Computational Linguistics* (ACL Anthology, 2019), doi:10.18653/v1/P19-1355.

51 **custom Tensor Processing Unit:** We tend to forget that our CPUs and GPUs are, for the most part, generalist machines. It's perfectly reasonable to build silicon devices for much more specific purposes, like matrix multiplication. Google's Tensor Processing Unit was first announced in 2016 and has undergone several generational updates since then.

52 **called the "Noösphere":** Herein is an interesting story of a terminology and worldview-that-could-have-been but got made complicated by the people involved. A nice brief review is by David Christian, who was answering *Edge* magazine's question of 2017 : "What scientific term or concept ought to be more widely known?" www.edge.org/response-detail/27068.

52 **in 1945 he wrote:** W. I. Vernadsky, "The Biosphere and the Noösphere," *American Scientist* 33, no. 1 (January 1945): xxii, 1–12, www.jstor.org/stable/27826043.

54 **Vaclav Smil:** He has written a book: Vaclav Smil, *Growth: From Microorganisms to Megacities* (Cambridge, MA: The MIT Press, 2019).

54 **Herman Hollerith:** He earned his Engineer of Mines (EM) degree from the Columbia University School of Mines in New York. His thinking on tabulating machines was part of his professional work; see, for example: Herman Hollerith, "An Electric Tabulating System," *The School of Mines Quarterly* 10, no.16 (April 1889): 238–55, www.columbia.edu/cu/computinghistory/hh/index.html.

55 **IBM was churning out:** It wasn't the sole producer; the RAND Corporation was the main competitor. A nice little history is given by IBM itself on its website: "The IBM Punched Card," www.ibm.com/ibm/history/ibm100/us/en/icons/punchcard/,

including the following lovely anecdote: "The phrase 'Do Not Fold, Spindle, or Mutilate' associated with punch cards inspired a 1971 movie starring Helen Hayes, Mildred Natwick, Myrna Loy and Sylvia Sidney as four elderly pranksters devoted to practical jokes." The title of the movie was *Do Not Fold, Spindle or Mutilate.* That phrase was also used in the Berkeley Free Speech Movement in the 1960s www.roughtype.com/?p=3182.

55 **200 *billion* cards:** I quote this figure and some others from a great article by George Dyson: "The Undead," *Wired* 7, no. 3 (March 1999), www.wired.com/1999/03/punchcards/.

55 **Margaret Hamilton:** She didn't just help get astronauts onto the Moon; she is widely credited with starting the use of the term "software engineering."

56 **coal-burning energy budget:** I've made this estimate by using approximate numbers from various sources on both US energy profiles over time and paper production, along with some "best guesses" at factors like transportation and human effort. I think it's actually likely that this underestimates the total effective energy burden.

56 **electromechanical organism:** Indeed, the punch card became one of the symbols of what the counter-culture movement railed against in the 1960s. See, for example: Robert MacBride, *The Automated State: Computer Systems as a New Force in Society* (Philadelphia: Chilton Book Co., 1967).

## 3. IN SICKNESS AND IN HEALTH

61 **as far back as 40,000 years:** Dating pictures like these is far from easy. In the case of the caves near Maros, an isotopic series of uranium was used in samples from the wonderfully named coralloid speleothems (or "cave popcorn," knobbly clusters of water-deposited calcite on top of the paintings and stencils): M. Aubert et al., "Pleistocene Cave Art from Sulawesi, Indonesia," *Nature* 514 (October 2014): 223–27, doi:10.1038/nature13422.

62 **65,000 years ago *Homo neanderthalensis*:** Not surprisingly, the evidence for art by other hominins is still debated and to some extent circumstantial—hinging on the timing of events and whether or not anatomically modern humans were doing this very early or whether Neanderthals had started before. See for example: A. W. G. Pike et al., "U-Series Dating of Paleolithic Art in 11 Caves in Spain," *Science* 336, no. 6087 (June 2012): 1409–13, doi:10.1126/science.1219957.

64 **a spiral of death:** Thanks to our online dataome, you can easily watch videos of army ant mills. They are quite something to see. These kinds of patterns are not unique to the ants, though; in fact the phenomenon of "vortex behaviors" in animal species is well-documented, albeit not fully understood. See for example: Johann Delcourt, Nikolai W. F. Bode, and Mathieu Denoël, "Collective Vortex Behaviors: Diversity, Proximate, and Ultimate Causes of Circular Animal Group Movements," *The Quarterly Review of Biology* 91, no. 1 (March 2016): 1–24, doi:10.1086/685301.

65 **label is "Big Data,":** As described in the text, pinning down the origins of the use of this term in the context of computation and statistics is not easy. A nice exploration was given in a piece in *The New York Times*: Steve Lohr, "The Origins of 'Big Data': An Etymological Detective Story," *Bits* [blog], February 1, 2013, https://bits.blogs.nytimes.com/2013/02/01/the-origins-of-big-data-an-etymological-detective-story/, in which the computer scientist John Mashey is pointed out as a likely culprit when he was chief scientist at Silicon Graphics in the 1990s.

66 **figures from the World Bank:** I'm sure there are many other sources, but the World Bank does keep a lot of handy and informative data online and freely available: https://data.worldbank.org/indicator/it.net.user.zs provides a graphic of internet use.

68 **"jpeg" or "gif":** Both of these terms, now part of common language, started as descriptors of computer file types ("image.jpeg," "image.gif"). The term "jpeg" or JPEG is an acronym for Joint Photographic Experts Group—a committee that decreed this discrete cosine transform technique as a standard in 1992, although the method was around since at least 1974. GIF refers to Graphics Interchange Format, another attempt at a standardized way of storing images, developed in the 1980s and utilizing a different kind of compression that encodes without loss.

68 **Compression is about eliminating extraneous information:** Of course, long before we used the term "compression" in the context of electronic data we practiced its use. Language compresses concepts into sounds, symbols, or gestures. Our brains likely also perform compression of a sort—not necessarily storing events or information in explicit form, but in ways that enable us to reconstruct things and fill in the blanks.

70 **very first transatlantic telegraph:** This cable went from Heart's Content in eastern Newfoundland to Valentia Island off the western coast of Ireland. The project belonged to the Atlantic Telegraph Company, founded just two years earlier.

70 **diplomatically choice words:** You can go read these for yourself, along with the transcript of Buchanan's response, online at the Library of Congress: www.loc.gov/pictures/item/2005694829/.

71 **dramatically—in the 1970s:** An interesting historical perspective on this is from the American Institute of Physics, which has developed a lesson plan for "'The Black Scientific Renaissance of the 1970s–90s:' African American Scientists at Bell Laboratories," https://www.aip.org/history-programs/physics-history/teaching-guides-women-minorities/black-scientific-renaissance-1970s-90s-african.

71 **electronic engineer Harry Nyquist:** Swedish-born Nyquist made one of the first uses of "information" as a term in the context of data transmission: Harry Nyquist, "Certain Factors Affecting Telegraph Speed," *The Bell System Technical Journal* 3, no. 2 (April 1924): 324–46, https://ieeexplore.ieee.org/document/6534511.

71 **"Transmission of Information.":** R. V. L. Hartley, "The Transmission of Information," *The Bell System Technical Journal* 7, no. 3 (July 1928): 535–63, doi:10.1002/j.1538-7305.1928.tb01236.x.

72 **"A Mathematical Theory of Communication,":** Despite its length and technical nature, this is a surprisingly fun read (well, perhaps it depends on your tastes):

C. E. Shannon, "A Mathematical Theory of Communication," *The Bell System Technical Journal* 27, no. 3 (July 1948): 379–423, doi:10.1002/j.1538-7305.1948.tb01338.x.

72 **special kind of character:** A terrific biography of Shannon is by Jimmy Soni and Rob Goodman, *A Mind at Play: How Claude Shannon Invented the Information Age* (New York: Simon and Schuster, 2017). Also see the article by Siobhan Roberts, "Claude Shannon, the Father of the Information Age, Turns 1100100," *The New Yorker*, April 30, 2016, www.newyorker.com/tech/annals-of-technology/claude-shannon-the-father-of-the-information-age-turns-1100100.

73 **call Boolean algebra:** The algebra of true and false (or 1 and 0) was first put forth by the mathematician George Boole in 1847 and properly in a book in 1854, just ten years before his death at only forty-nine after getting pneumonia. At its core Boolean algebra involves the logical operations of AND, OR, and NOT. For example: *x* AND *y* is only true (or 1) if both *x* and *y* are true. So the algebra lends itself to electronic circuits where switches can be on or off (true or false) and circuits can be configured to produce AND, OR, and NOT current flows.

73 **to his chagrin:** Although he expressed it gently, his feelings are pretty evident in this 1956 editorial: Claude E. Shannon, "The Bandwagon," *IRE Transactions on Information Theory* 2, no. 1 (March 1956): 3, doi:10.1109/TIT.1956.1056774.

74 **now ubiquitous "bit.":** To quote Shannon exactly from his 1948 paper: "If the base 2 is used the resulting units may be called binary digits, or more briefly bits, a word suggested by J. W. Tukey."

77 **Our precious letters and words:** There is a large and fascinating subfield of "natural language processing" that concerns itself with both the nature of machines and language and the intrinsic structure of language itself (irrespective of which language). This ranges from the rules of language to the statistics of language. But there is also linguistics as developed by the likes of the remarkable Noam Chomsky, with theoretical, mathematical constructs that attempt to connect language to evolutionary biology.

77 **yielded a formula:** If only this were easier for most of us to understand immediately. There is a large literature—both instructional and in research—on Shannon's formula. Very little of it produces any kind of "aha!" moment of insight. You have to struggle a bit to think it through. By giving a few different examples and approaches here, I've tried to crack the door open for the reader. One example of an extremely nice, albeit technical, introduction to information theory is by the polymath Simon DeDeo: "Information Theory for Intelligent People," 2018, http://tuvalu.santafe.edu/~simon/it.pdf. In case you're wondering, the formula can be written as: $H = -\sum_i P_i \log P_i$. This mathematical function adds up the product of two quantities, a probability **P** and the logarithm of that probability **log P**. The probability is how likely a given symbol is to occur in a dataset—whether a letter in a sentence, a word, or anything else. By using a logarithm Shannon could conveniently sum up rather than multiply all the probabilities together. If that logarithm uses base 2 (as opposed to our common, day-to-day use of base 10), it takes

on an interesting property. Think back to our coin with its 50/50 chance of turning up heads or tails. The base 2 logarithm of 0.5 (50 percent) is –1. So, the summation of Shannon's formula for **H** for an unbiased coin (which can be either heads or tails) is –(–0.5–0.5), which equals 1, or one bit of information. But if the coin is biased, then **H** is always less than 1. Just like before, the biased coin always stores less than one bit of information.

79 **most widely told anecdote:** There are numerous retellings and versions of this. A pretty convincing one is from a conversation between Myron Tribus (an engineer and thermodynamicist) and Shannon that purportedly took place in 1961: see Myron Tribus, "Information Theory and Thermodynamics," in: *Heat Transfer, Thermodynamics and Education*: Boelter Anniversary Volume, ed. Harold A. Johnson and Llewellyn M. K. Boelter, 348–68 (New York: McGraw-Hill: 1964), 354.

82 **Carlo Rovelli, writing in 2016:** Rovelli's short article "Meaning = Information + Evolution" (an online preprint only, https://arxiv.org/abs/1611.02420) is a very nice review-level piece, and explicitly makes use of a simple model proposed by David H. Wolpert and Artemy Kolchinsky: "Observers as Systems That Acquire Information to Stay out of Equilibrium," in Foundational Questions Institute (FQXi) 5th Annual Conference, "If a Tree Falls: The Physics of What Happens and Who Is Listening?" (uniting "The Physics of the Observer" and "The Physics of What Happens"), Banff, Alberta, Canada, August 17–22, 2016.

85 **redundancy in its root coding:** As I've written, our first guess is that this provides some resilience against point mutations. But it may be more than just that. Different molecular codes for the same thing may also produce different translation rates—how fast a ribosome will turn code into a protein. This could play a role in ensuring correctly folded end proteins, for example.

87 **eye-opening study:** Martin Hilbert and Priscila López, "The World's Technological Capacity to Store, Communicate, and Compute Information," *Science* 332, no. 6025 (April 2011): 60–65, doi:10.1126/science.1200970.

88 **crested over 2,000 exabytes:** One exabyte is a billion gigabytes. These numbers are always estimates, and can be found at various sources, often targeting commerce, for example: www.statista.com/statistics/638593/worldwide-data-center-storage-capacity-cloud-vs-traditional/ or www.weforum.org/agenda/2019/04/how-much-data-is-generated-each-day-cf4bddf29f/.

91 ***E. coli,* for instance:** The trickery and strategies of microbes are something to behold. A short, readable review of some of these is: Erica Bizzell, "Microbial Ninja Warriors: Bacterial Immune Evasion," American Society for Microbiology, December 2018, www.asm.org/Articles/2018/December/Microbial-Ninja-Warriors-Bacterial-Immune-Evasion and references therein.

91 **spider-tailed horned viper:** This is honestly one of the most bizarre and extraordinary creatures. It's also an example of natural selection "encoding" information about the external world in the traits of an organism. It is endemic to western Iran, and if you go online you can find some remarkable videos of the snake in action, luring and catching birds.

94 **gray areas too:** There are plenty of these when it comes to sexual selection. See, for example: Suzanne H. Alonzo and Maria R. Servedio, "Grey Zones of Sexual Selection: Why Is Finding a Modern Definition So Hard?," *Proceedings of the Royal Society B* 286, no. 1909 (August 2019): 20191325, doi:10.1098/rspb.2019.1325.

94 **carried out in 2013:** Morgan David et al., "Pairing Context Determines Condition-Dependence of Song Rate in a Monogamous Passerine Bird," *Proceedings of the Royal Society B* 280, no. 1753 (February 2013): 20122177, doi:10.1098/rspb.2012.2177.

95 **Richard Prum posited:** These ideas were gathered in a book: Richard O. Prum, *The Evolution of Beauty: How Darwin's Forgotten Theory of Mate Choice Shapes the Animal World—and Us* (New York: Doubleday, 2017). But see also some of the commentary and critique, such as that by Douglas J. Futuyma in *The Quarterly Review of Biology* 93, no. 2 (June 2018): 150–51.

96 **Generative Adversarial Networks (GANS):** are a comparatively recent invention. See: Ian J. Goodfellow et al., "Generative Adversarial Nets," NIPS 2014: Proceedings of the 27th International Conference on Advances in Neural Information Processing Systems, vol. 2 (December 2014): 2672–80, https://papers.nips.cc/paper/5423-generative-adversarial-nets.pdf. The term "generative" refers to a distinct statistical modeling approach.

98 **quite easily fooled:** While this is not news to researchers at the forefront of machine learning, you'd sometimes be hard-pressed to notice such facts in popular media or commercial enterprises. See, for example: Douglas Heaven, "Why Deep-Learning AIs Are So Easy to Fool," *Nature* 574 (October 2019): 163–66, doi:10.1038/d41586-019-03013-5.

99 **"bit rot.":** While the name of internet pioneer Vint Cerf is generally associated with this term, it's not entirely clear when it originated. But Cerf himself has talked about the problem on many occasions. See, for example: Adam Chandler, "A Warehouse Fire of Digital Memories," *The Atlantic,* February 13, 2015, www.theatlantic.com/technology/archive/2015/02/google-forgotten-century-digital-files-bit-rot/385500/.

## 4. AN EVER MORE TANGLED BANK

101 **James Baldwin (1963):** What an extraordinary mind. This quote comes from an article by Jane Howard, "Doom and Glory of Knowing Who You Are," *LIFE* 54, no. 21 (May 24, 1963): 89.

**102 his *On the Origin of Species*:** The most complete and readily available Darwin resource is, naturally, online at Darwin Online, with over 219,000 pages of searchable text and one of the largest collections of Darwin's private papers: http://darwin-online.org.uk/, cited as John van Wyhe, ed., 2002, The Complete Work of Charles Darwin Online.

**103 *grandeur in this view*:** In the interests of brevity I left out the text immediately prior to this sentence. To fill in the gap, here is that text from *On the Origin of*

*Species* (1st edition): "These laws, taken in the largest sense, being Growth with Reproduction; Inheritance which is almost implied by reproduction; Variability from the indirect and direct action of the external conditions of life, and from use and disuse; a Ratio of Increase so high as to lead to a Struggle for Life, and as a consequence to Natural Selection, entailing Divergence of Character and the Extinction of less-improved forms. Thus, from the war of nature, from famine and death, the most exalted object which we are capable of conceiving, namely, the production of the higher animals, directly follows."

104 **some 177 neurons:** A nice review and overview of small brains is: Luis A. Bezares-Calderón and Gáspár Jékely, "Think Small," *eLife* 5 (December 2016): e22497, doi:10.7554/eLife.22497.

105 **Sea sponges:** This is quite a controversial topic (yes, this is what scientists spend their time worrying about). See, for example: Frank Hirth, "On the Origin and Evolution of the Tripartite Brain," *Brain, Behavior and Evolution* 76, no. 1 (October 2010): 3–10, doi:10.1159/000320218. Also a popular-level article: Cara Giaimo, "A Battle Is Raging in the Tree of Life," *The New York Times,* August 2, 2019, www.nytimes.com/2019/08/02/science/tree-of-life-sponges-jellies.html.

105 **brain size versus body size:** Often discussed in terms of an "encephalization quotient"; see, for example: Osvaldo Cairó, "External Measures of Cognition," *Frontiers in Human Neuroscience* 5 (October 2011): 108, doi:10.3389/fnhum.2011.00108.

106 **bony-eared assfish:** A species that is likely utterly incapable of ever knowing that another species snickers about its name in one particular language. It's a cool little organism though, and may have really good hearing: M. L. Fine, M. H. Horn, and Brian Cox, "*Acanthonus armatus,* a Deep-Sea Teleost Fish with a Minute Brain and Large Ears," *Proceedings of the Royal Society B* 230, no. 1259 (March 1987): 257–65, doi:10.1098/rspb.1987.0018.

106 **metabolic activity of a nervous system:** There seem to be relatively few sources in which data has been collated across different species. One slightly old version is: J. W. Mink, R. J. Blumenschine, and D. B. Adams, "Ratio of Central Nervous System to Body Metabolism in Vertebrates: Its Constancy and Functional Basis," *American Journal of Physiology* 241, no. 3 (September 1981): R203–12, doi:10.1152/ajpregu.1981.241.3.R203.

107 **edge of self-organized criticality:** See, for example: Antonio J. Fontenele et al., "Criticality between Cortical States," *Physical Review Letters* 122 (May 2019): 208101, doi:10.1103/PhysRevLett.122.208101; and a popular-level article: Charlie Wood, "Do Brains Operate at a Tipping Point? New Clues and Complications," *Quanta Magazine*, June 10, 2019, www.quantamagazine.org/do-brains-operate-at-a-tipping-point-new-clues-and-complications-20190610/. For a clear explanation of criticality see: Jennifer Ouellette, "A Fundamental Theory to Model the Mind," *Quanta Magazine, April 3, 2014,* https://www.quantamagazine.org/toward-a-theory-of-self-organized-criticality-in-the-brain-20140403/.

107 ***The Energies of Men***: James seems to have variously given this material as a speech and as a paper in more than one place, but here is one accessible reference: William

James, "The Energies of Men," *The Philosophical Review* 16, no. 1 (January 1907): 1–20, doi:10.2307/2177575.

108 **full of variability and noise:** A nice popular-level piece is: Michael Segal, "Why the Brain Is So Noisy," *Nautilus* 68, January 17, 2019, http://nautil.us/issue/68/context/why-the-brain-is-so-noisy.

109 **study presented in 2019:** A universal transmission rate for natural human communication is pretty astonishing. The research paper is: Christophe Coupé et al., "Different Languages, Similar Encoding Efficiency: Comparable Information Rates across the Human Communicative Niche," *Science Advances* 5, no. 9 (September 2019): eaaw2594, doi:10.1126/sciadv.aaw2594.

110 **only as efficient as necessary:** At the same time, this certainly doesn't mean that brains are inefficient—it's all relative. See, for example, the discussion and comparison to digital computation in Liqun Luo, "Why Is the Human Brain So Efficient?," *Nautilus* 59, April 12, 2018, http://nautil.us/issue/59/connections/why-is-the-human-brain-so-efficient.

110 **A research study:** The paper is Michael A. Skeide et al., "Learning to Read Alters Cortico-Subcortical Cross-Talk in the Visual System of Illiterates," *Science Advances* 3, no. 5 (May 2017): e1602612, doi:10.1126/sciadv.1602612.

111 **"neuronal recycling":** Proposed in Stanislas Dehaene and Laurent Cohen, "Cultural Recycling of Cortical Maps," *Neuron* 56, no. 2 (October 2007): 384–98, doi:10.1016/j.neuron.2007.10.004.

113 **called intrinsic forgetting:** An excellent paper that also introduced the concept is Ronald L. Davis and Yi Zhong, "The Biology of Forgetting—A Perspective," *Neuron* 95, no. 3 (August 2017): 490–503, doi:10.1016/j.neuron.2017.05.039.

113 **unfortunate sea slugs:** Their genus name is *Aplysia* and they've been a favorite for studies of physical neural changes because they have only about 20,000 neurons and their gill- and siphon-withdrawal reflex is nicely associated with specific neurons. And presumably they don't complain about the indignities and terrors of being dissected and examined by humans. One example of the retention of memory is Shanping Chen et al., "Reinstatement of Long-Term Memory Following Erasure of Its Behavioral and Synaptic Expression in Aplysia," *eLife* 3 (November 2014): e03896, doi:10.7554/eLife.03896.

114 **the Antikythera mechanism:** This object was brought up from a 2,000-year-old shipwreck off the Greek island of Antikythera in 1902, inside the remains of a wooden box. Modern x-ray tomography and microscopy has enabled researchers to deconstruct what was a corroded, fused chunk of bronze. It contains 29 gears, indications of ten more gears, plus pointers, shafts, and axles, that seem to correspond to what is, in effect, a computational machine for predicting lunar and solar motions. See for example, Kyriakos Efstathiou and Marianna Efstathiou, "Celestial Gearbox," *Mechanical Engineering* 140, no. 9 (September 2018): 31–35, doi:10.1115/1.2018-SEP1.

115 **tenth-century appendix:** See also Bruce Eastwood and Gerd Graßhoff, "Planetary Diagrams for Roman Astronomy in Medieval Europe, ca. 800–1500,"

*Transactions of the American Philosophical Society, n.s.,* 94, no. 3 (2004), doi:10.2307/20020363.

116 **"moral statistics" of France:** The essay was A-M. Guerry, "Essai sur la Statistique Morale de la France," a slender set of pages filled with figures and tables; it was groundbreaking. See for example: Michael Friendly, "A.-M. Guerry's *Moral Statistics of France*: Challenges for Multivariable Spatial Analysis," *Statistical Science* 22, no. 3 (2007): 368–99, doi:10.1214/07-STS241. Also a great resource with some of Guerry's original map images is: http://datavis.ca/gallery/guerry/.

118 **watching videos together:** It can be tempting to giggle at this kind of research, but in truth it's remarkably informative and fascinating: Wouter Wolf and Michael Tomasello, "Visually Attending to a Video Together Facilitates Great Ape Social Closeness," *Proceedings of the Royal Society B* 286, no. 1907 (July 2019): 20190488, doi:10.1098/rspb.2019.0488.

118 **2017 Gojko Barjamovic:** I got so excited when I came across this work; it's just lovely. The original report was as a working paper in 2017, but it more recently got fully published: Gojko Barjamovic et al., "Trade, Merchants, and the Lost Cities of the Bronze Age," *The Quarterly Journal of Economics* 134, no. 3 (August 2019): 1455–1503, doi:10.1093/qje/qjz009.

121 **projects like "Old Weather":** Look past the near-oxymoron of the name and you'll find a fantastic project that you can participate in, if you like poring over old documents written by crusty old sea dogs (sorry, not really, just dutiful captains and their crew): https://www.oldweather.org/.

121 **on the city of Venice:** This effort has run into some controversy and problems, and at the time of writing is in limbo, but it remains a good example of the possibilities: Alison Abbott, "The 'Time Machine' Reconstructing Ancient Venice's Social Networks," *Nature* 546 (June 2017): 341–44, doi:10.1038/546341a; and an update on its woes: Davide Castelvecchi, "Venice 'Time Machine' Project Suspended amid Data Row," *Nature* 574 (October 2019): 607, doi:10.1038/d41586-019-03240-w.

122 **thoughts and observations:** Again, the Darwin Online resource has much on this: http://darwin-online.org.uk/EditorialIntroductions/Chancellor_GeologicalDiary.html. I was also made aware of this tendency of Darwin's in a casual conversation with the historian of science Luis Campos, who is a professor at The University of New Mexico.

123 **three genetic mutations:** See for example: Marianthi Karageorgi et al. "Genome Editing Retraces the Evolution of Toxin Resistance in the Monarch Butterfly," *Nature* 574 (October 2019): 409–12, doi:10.1038/s41586-019-1610-8; and also: Andrew M. Taverner et al., "Adaptive Substitutions Underlying Cardiac Glycoside Insensitivity in Insects Exhibit Epistasis In Vivo," *eLife* 8 (August 2019): e48224, doi:10.7554/eLife.48224.

124 **The scientist Ilan Eshel:** See, for example: Ilan Eshel, "Clone Selection and the Evolution of Modifying Features," *Theoretical Population Biology* 4, no. 2 (June 1973): 196–208, doi:10.1016/0040-5809(73)90029-4.

126 **a booklet circulates:** A fact reported in various places, see for example: Yan Jiaqi and Gao Gao, *Turbulent Decade: A History of the Cultural Revolution*, 1st ed. (Honolulu: University of Hawai'i Press, 1996), 401–2.

128 **Krakauer beautifully articulated:** The paper is David C. Krakauer, "Darwinian Demons, Evolutionary Complexity, and Information Maximization," *Chaos* 21 (September 2011): 037110, doi.10.1063/1.3643064, and it will crop up again later.

## 5. GENES, MEMES, AND DREAMS

131 **Ada Lovelace, 1841:** This quotation is from a letter reproduced in the book edited by Betty A. Toole *Ada, the Enchantress of Numbers: A Selection from the Letters of Lord Byron's Daughter and Her Description of the First Computer* (London: The Strawberry Press, 1992).

135 **called Hamilton's Rule:** The original papers on this were: W. D. Hamilton, "The Genetical Evolution of Social Behaviour. I," *Journal of Theoretical Biology* 7, no. 1 (July 1964): 1–16, https://doi.org/10.1016/0022-5193(64)90038-4; and Hamilton, "The Genetical Evolution of Social Behaviour. II," *Journal of Theoretical Biology* 7, no. 1 (July 1964): 17–52 https://doi.org/10.1016/0022-5193(64)90039-6, and while the end general rule is pretty simple-looking, the mathematics and arguments are not for the fainthearted. Also, a very nice popular-level piece on the ongoing struggle to resolve some of the puzzles behind these ideas: Jordana Cepelewicz, "The Elusive Calculus of Insects' Altruism and Kin Selection," *Quanta Magazine*, April 10, 2018, www.quantamagazine.org/the-elusive-calculus-of-insects-altruism-and-kin-selection-20180410/.

136 **science book *The Selfish Gene*:** For the record: Richard Dawkins, *The Selfish Gene* (New York: Oxford University Press, 1976).

136 **outsized amount of debate:** It's funny how this book, and Dawkins's ideas, are seen by professional biologists and evolutionary biologists. Most of my conversations with these kinds of people end up with them telling me that the notion of the "selfish" gene is sometimes helpful but far from the whole picture. That makes sense to me. At the same time, as a physicist I am drawn to the extreme reductionism of the idea.

136 **and the biologist George Williams:** Hamilton was an English evolutionary biologist famous for the eponymous "Hamilton's Rule," a theoretical framework that can be used to try to explain individual and social behavior as consequences of a cost-benefit relationship and the propagation of shared genes in related organisms. George C. Williams was an American evolutionary biologist who made major contributions to how we see the mechanisms of selection and evolution—including a shift to seeing "adaptation" as a consequence of selection, not some kind of active mechanism of survival.

137 **groups and societies:** And herein lies the tale of contentious arguments about distinctions that are to some extent hard for outsiders to grasp. Although it is all

fascinating, I hesitate to say too much, except that an example of a pro–group fitness argument is to be found in Martin A. Nowak, Corina E. Tarnita, and Edward O. Wilson, "The Evolution of Eusociality," *Nature* 466 (August 2010): 1057–62, doi:10.1038/nature09205. I personally find that the idea of multilevel selection (which blends things together from genes to groups and beyond) seems to make sense. A nice academic review of the tension between some of these ideas is: Jonathan Birch, "Are Kin and Group Selection Rivals or Friends?," *Current Biology* 29, no. 11 (June 2019): R433–38, doi:10.1016/j.cub.2019.01.065.

137 **We know of "supergenes,":** This is really bizarre and also really cool. The example of the otherwise innocent-sounding ruff, a cute little wading bird, is wild: Clemens Küpper et al., "A Supergene Determines Highly Divergent Male Reproductive Morphs in the Ruff," *Nature Genetics* 48 (2016): 79–83, doi:10.1038/ng.3443.

138 **primates and humans:** For example, see the study by Alastair Crisp et al., "Expression of Multiple Horizontally Acquired Genes Is a Hallmark of Both Vertebrate and Invertebrate Genomes," *Genome Biology* 16 (March 2015): 50, doi:10.1186/s13059-015-0607-3.

138 **there are transcription factors:** In many ways I'm understating the likely importance of these proteins and their role not just in the operational aspects of an organism but also in evolution itself. They take us even further from the old notion of DNA as some kind of "hard drive" or "blueprint" and toward a much more dynamic picture, where DNA is a highly compressed dataset that can be utilized on the fly. See, for example: Samuel A. Lambert et al., "The Human Transcription Factors," *Cell* 172, no. 4 (February 2018): 650–65, doi:10.1016/j.cell.2018.01.029.

139 **by its rivals:** In an essay: T. H. Huxley, "The Coming of Age of the Origin of Species," *Nature* 22 (May 1880): 1–4, doi:10.1038/022001a0.

142 **are called 16S rRNA:** Some of the work pioneering the use of 16S rRNA for phylogenetic studies was done in the 1970s by Carl Woese and George Fox, leading to the identification of an entire "other" domain of life: the archaea.

143 **the chemist Julius Rebek:** For example see his review: Julius Rebek Jr., "Molecular Recognition and Self-Replication," *Journal of Molecular Recognition* 5, no. 3 (September 1992): 83–88, doi:10.1002/jmr.300050302.

143 **Douglas Hofstadter in 1979:** In his most famous book: Douglas R. Hofstadter, *Gödel, Escher, Bach: An Eternal Golden Braid* (New York: Basic Books, 1979).

144 **these "de novo" genes:** A good review is: Adam Levy, "How Evolution Builds Genes from Scratch," *Nature* 574 (October 2019): 314–16, doi:10.1038/d41586-019-03061-x. It is also clear that these genes are only "from scratch" inasmuch as they're like a new mixtape (sorry, Spotify playlist), using sequences that may have at some point been active or incorporated from other places.

146 **uses metaphor and analogy:** I'll unashamedly refer the reader to a piece I wrote about this: Caleb A. Scharf, "In Defense of Metaphors in Science Writing," *Scientific American*, July 9, 2013, https://blogs.scientificamerican.com/life-unbounded/in-defense-of-metaphors-in-science-writing/.

147 **Barbara McClintock once said:** Quoted in Evelyn Fox Keller, *A Feeling for the Organism: The Life and Work of Barbara McClintock* (New York: Henry Holt and Company, 1983), 200.

147 **an ongoing debate:** See, for example: John W. Pepper and Matthew D. Herron, "Does Biology Need an Organism Concept?," *Biological Reviews* 83, no. 4 (November 2008): 621–27, doi:10.1111/j.1469-185X.2008.00057.x.

147 **promoted the "restoration":** This came up in his paper: Stephen Jay Gould, "Is a New and General Theory of Evolution Emerging?," *Paleobiology* 6, no. 1 (Winter 1980): 119–30, www.jstor.org/stable/2400240. And to quote from the abstract of this paper: "A new and general evolutionary theory will embody this notion of hierarchy and stress a variety of themes either ignored or explicitly rejected by the modern synthesis: punctuational change at all levels, important non-adaptive change at all levels, control of evolution not only by selection, but equally by constraints of history, development and architecture—thus restoring to evolutionary theory a concept of organism."

148 **Immanuel Kant in 1781:** The notion of the transcendental is in Kant's *Critique of Pure Reason* (*Kritik der reinen Vernunft*), published in Germany, and in later writings.

148 **a provocative research paper:** This is: Scott F. Gilbert, Jan Sapp, and Alfred I. Tauber, "A Symbiotic View of Life: We Have Never Been Individuals," *The Quarterly Review of Biology* 87, no. 4 (December 2012): 325–41, doi:10.1086/668166.

149 **That implies several things:** The exact term *holobiont* is generally credited to the evolutionary biologist Lynn Margulis, who used it in the early 1990s. The idea of "holobiosis" was also around much earlier, in the 1940s, due to the theoretical biologist Adolf Meyer-Abich. See for example: Jan Baedke, Alejandro Fábregas-Tejeda, and Abigail Nieves Delgado, "The Holobiont Concept before Margulis," *Journal of Experiental Zoology* (*Molecular and Developmental Evolution*) 334, no. 3 (May 2020): 149–55, doi:10.1002/jez.b.22931.

150 ***ecosystem* than as a holobiont:** There are some pretty vigorous arguments going on about this. A further review and overview is: J. Jeffrey Morris, "What Is the Hologenome Concept of Evolution?," *F1000 Research* 7 (F1000 Faculty Rev, October 2018): 1664, doi:10.12688/f1000research.14385.1; and also: Angela E. Douglas and John H. Werren, "Holes in the Hologenome: Why Host-Microbe Symbioses Are Not Holobionts," *mBio* 7, no. 2 (March 2016): e02099-15, doi:10.1128/mBio.02099-15.

150 **"It's the song, not the singer.":** What a brilliant title: W. Ford Doolittle and Austin Booth, "It's the Song, Not the Singer: An Exploration of Holobiosis and Evolutionary Theory," *Biology and Philosophy* 32, no. 1 (January 2017): 5–24, doi:10.1007/s10539-016-9542-2.

152 **throughout life's history:** A very nice technical argument for this is: Paul G. Falkowski, Tom Fenchel, and Edward F. Delong, "The Microbial Engines that Drive Earth's Biogeochemical Cycles," *Science* 320, no. 5879 (May 2008): 1034–39, doi:10.1126/science.1153213.

153 **or MC1R gene:** And there are interactions between the products of these genes; see, for example: Michael M. Ollmann et al., "Interaction of Agouti Protein with the Melanocortin 1 Receptor In Vitro and In Vivo," *Genes & Development* 12, no. 3 (February 1998): 316–30, doi:10.1101/gad.12.3.316.

155 **a Saturn V rocket:** Although enormously complex, the Saturn V, which NASA built to send people to the Moon, may actually not have been as complex as something like NASA's Space Shuttle, which had over 2.5 million distinct parts, including 370 kilometers (230 miles) of wiring and 1,440 circuit breakers.

156 **Gian-Carlo Rota:** It's a hefty read, but it is fascinating and lively too: Gian-Carlo Rota, "Lectures on Being and Time (1998)," *The New Yearbook for Phenomenology and Phenomenological Philosophy* VIII (2008), 225–319.

157 ***Category theory* is a theory:** I'd like to thank Dr. Will Cavendish for helping demystify some of category theory during a lively group discussion at the Institute for Advanced Study in Princeton, on a wet day in March 2020.

158 **divided between impressions and ideas:** Something still mulled over today; see, for example: Samuel C. Rickless, "Hume's Distinction between Impressions and Ideas," *European Journal of Philosophy* 26, no. 4 (December 2018): 1222–37, doi:10.1111/ejop.12347.

158 **Leibniz in the 1660s:** His "Dissertation on the Art of Combinations" can today be found in G. W. Leibniz, *Philosophical Papers and Letters,* ed. Leroy E. Loemker, 2nd ed., 73–84, *The New Synthese Historical Library: Texts and Studies in the History of Philosophy*, vol. 2. (Dordrecht, The Netherlands: Kluwer Academic Publishers, 1989).

158 **called Abraham Abulafia:** See, for example: Joseph Dan, *The Heart and the Fountain: An Anthology of Jewish Mystical Experiences* (New York: Oxford University Press, 2002). Also: John F. Nash, "Abraham Abulafia and the Ecstatic Kabbalah," *The Esoteric Quarterly 4,* no. 3 (Fall 2008): 51–64.

158 **NLP, *natural language processing*:** The most basic example of language processing is about predicting what the next word will be in a sentence. But beyond that is the challenge of making systems that can both parse and generate natural, human-sounding language. We'll come back to this later in talking about Alan Turing.

159 **it was good stuff:** There are many sources on Leibniz to look at, but a nice compact one is: Maria Rosa Antognazza, *Leibniz: A Very Short Introduction* (New York: Oxford University Press, 2016).

159 **essay written in 2017:** The text is exclusively online: Kenneth O. Stanley, Joel Lehman, and Lisa Soros, "Open-Endedness: The Last Grand Challenge You've Never Heard Of," O'Reilly, December 19, 2017, www.oreilly.com/radar/open-endedness-the-last-grand-challenge-youve-never-heard-of/.

160 **an "information bomb":** This kind of discussion is in Richard Dawkins, *River out of Eden: A Darwinian View of Life* (New York: Basic Books, 1995).

161 **take a leap of imagination:** In machine learning people sometimes talk about having their systems "dream," or perform what's called domain randomization. But I don't think this should be conflated with the idea of genuine, neural imagination. Indeed, the procedurally generated content in machine "dreaming" is designed to

help with overfitting in artificial neural nets. See, for example: Niels Justesen et al., "Illuminating Generalization in Deep Reinforcement Learning through Procedural Level Generation," NeurIPS Deep RL Workshop 2018, https://arxiv.org/abs/1806.10729. Although, perhaps this is part of what we do when we dream—randomly generate scenarios to better assess memories and experiences. And maybe why fiction is so alluring: there is a selective advantage to a species playing out hypothetical scenarios.

162 **the "Fluctuation Theorem,":** Although the original work was in 1993/1994, this update is perhaps a better place to start: Denis J. Evans and Debra J. Searles, "The Fluctuation Theorem," *Advances in Physics* 51, no. 7 (2002): 1529–85, doi:10.1080/00018730210155133.

162 **scrub it and sterilize it:** On a related note, we don't really know what the details of a truly "abiotic" planet similar to Earth would be. If life had never happened here, there would be quite different environmental chemistry, different surface minerals, and so on.

164 **an information theory of individuality:** See David Krakauer et al., "The Information Theory of Individuality," *Theory in Biosciences* 139 (March 2020): 209–23, doi:10.1007/s12064-020-00313-7.

## 6. THE INFORMATION RIVER

168 **shared space to spread across:** This is not always mentioned, but I think it's the most intuitive way to think about covalent bonding (as much as anything in quantum mechanics is intuitive). See, for example: Michael W. Schmidt, Joseph Ivanic, and Klaus Ruedenberg, "Covalent Bonds Are Created by the Drive of Electron Waves to Lower Their Kinetic Energy through Expansion," *The Journal of Chemical Physics* 140, no. 20 (2014): 204104, doi:10.1063/1.4875735.

172 **how these signaling peptides:** For example, see: Zohar Erez et al., "Communication between Viruses Guides Lysis–Lysogeny Decisions," *Nature* 541 (January 2017): 488–93, doi:10.1038/nature21049.

172 **infecting soil bacteria:** See, for example: Avigail Stokar-Avihail et al., "Widespread Utilization of Peptide Communication in Phages Infecting Soil and Pathogenic Bacteria," *Cell Host & Microbe* 25, no. 5 (May 2019): 746–55.e5, doi:10.1016/j.chom.2019.03.017.

173 **can trigger interference:** See Hannah G. Hampton, Bridget N. J. Watson, and Peter C. Fineran, "The Arms Race between Bacteria and Their Phage Foes," *Nature* 577 (January 2020): 327–36, doi:10.1038/s41586-019-1894-8.

173 **Paul Turner and Lin Chao:** See Paul E. Turner and Lin Chao, "Prisoner's Dilemma in an RNA Virus," *Nature* 398 (April 1999): 441–43, doi:10.1038/18913.

174 **nascent field of "sociovirology":** See, for example: Elie Dolgin, "The Secret Social Lives of Viruses," *Nature* 570 (June 2019): 290–92, doi:10.1038/d41586-019-01880-6.

174 **exosomes in particular:** I hadn't appreciated any of this until reading this nice review paper: Raghu Kalluri and Valerie S. LeBleu, "The Biology, Function, and Biomedical Applications of Exosomes," *Science* 367, no. 6478 (February 2020): eaau6977, doi:10.1126/science.aau6977.

175 **Schrödinger made key predictions:** Specifically, he postulated an "aperiodic crystal" with covalent bonds as genetic material—a structure that could encode the heritable information of life. This was entirely based on theoretical needs for living systems, but it turned out to be remarkably prescient.

177 **Earth-Life Science Institute:** This center, known more commonly as ELSI, is a $100 million experiment, one of the visionary institutes supported by the Japanese government's World Premier International Research Center Initiative (WPI). And in full disclosure I have close ties with it through my role as a Global Science Coordinator for a project on the origins of life, run within ELSI. Many a happy day has been spent in ELSI's lovely collaborative spaces.

178 **casinos in Las Vegas:** Someone wrote a book about this: Thomas A. Bass, *The Eudaemonic Pie: The Bizarre True Story of How a Band of Physicists and Computer Wizards Took On Las Vegas* (Boston: Houghton Mifflin, 1985).

178 **life's deepest properties:** I'd be remiss not to point the reader to Eric's own coauthored book on life: Eric Smith and Harold J. Morowitz, *The Origins and Nature of Life on Earth: The Emergence of the Fourth Geosphere* (Cambridge, UK: Cambridge University Press, 2016).

179 **mathematician John Conway:** Who sadly passed away from complications of COVID-19 while I was writing this book, after a much-admired career at both the University of Cambridge and Princeton University and major contributions to mathematics that extended far from cellular automata (the popularity of which seemed to eventually grate on Conway).

179 **the scientist Stephen Wolfram:** has pursued cellular automata as an elementary piece of his ideas on "new physics" and framing the world through rules and computation. It is fascinating, but has also met with intense skepticism from the scientific community.

179 **simulations of "boids":** The origins of which are credited to Craig W. Reynolds, "Flocks, Herds and Schools: A Distributed Behavioral Model," *ACM SIGGRAPH Computer Graphics* 21, no. 4 (August 1987): 25–34, doi:10.1145/37401.37406.

182 **"informational narrative of living systems.":** See Sara Imari Walker and Paul C. W. Davies, "The Algorithmic Origins of Life," *Journal of the Royal Society Interface* 10, no. 79 (February 2013): 20120869, doi:10.1098/rsif.2012.0869.

183 **in a 2016 book, states:** George Ellis, *How Can Physics Underlie the Mind? Top-Down Causation in the Human Context* (Berlin and Heidelberg: Springer-Verlag, 2016).

186 **termed symmetry breaking:** A key part of symmetry breaking in physics is that the moment of "breaking" will take a system from more disorder to more order, but the more ordered state need not be easily predictable, since there can be many, many such states.

187 **complexity in nature?:** James P. Crutchfield, "The Calculi of Emergence: Computation, Dynamics and Induction," *Physica D: Nonlinear Phenomena* 75, nos. 1–3 (August 1994): 11–54, doi:10.1016/0167-2789(94)90273-9.

188 **Maxwell imagined a being:** It seems he first wrote about this in 1867 in a letter to the physicist Peter Tait, and eventually published the idea in his 1871 book on thermodynamics: J. Clerk Maxwell, *Theory of Heat* (New York: Longmans, Green, and Company, 1871). An online readable version is: https://archive.org/details/theoryofheat00maxwrich/page/n8/mode/2up.

189 **Published in 1929:** And here it is (in German): L. Szilard, "über die Entropieverminderung in einem thermodynamischen System bei Eingriffen intelligenter Wesen [On the Reduction of Entropy in a Thermodynamic System by the Intervention of Intelligent Beings]," *Zeitschrift für Physik* 53 (November 1929), 840–56, doi:10.1007/BF01341281.

190 **group of scientists in Japan:** The experiment is reported in a fascinating, if technical paper: Shoichi Toyabe et al., "Experimental Demonstration of Information-to-Energy Conversion and Validation of the Generalized Jarzynski Equality," *Nature Physics* 6 (November 2010): 988–92, doi:10.1038/nphys1821.

190 **Brownian ratchet:** in theory, a class of perpetual motion machine. The "Brownian" refers to the phenomenon of Brownian motion, the random motion of particles jostled by the molecules of a gas or liquid. The ratchet idea was due to the physicist Marian Smoluchowski in 1912 and was revived in 1962 by the physicist Richard Feynman, who also helped show why it's not a perpetual motion machine—because its pieces are being jiggled by random motion too, causing the ratchet action to fail as often as it succeeds (when everything is at the same temperature).

191 **Landauer's insight:** Rolf Landauer, "Irreversibility and Heat Generation in the Computing Process," *IBM Journal of Research and Development 5, no. 3* (July 1961): 183–91, doi:10.1147/rd.53.0183.

192 **Experimental work in 2018:** See L. L. Yan et al., "Single-Atom Demonstration of the Quantum Landauer Principle," *Physical Review Letters* 120, no. 21 (May 2018): 210601, doi:10.1103/PhysRevLett.120.210601.

193 **"Darwinian Demon":** I referenced this paper before, and here we are again. Krakauer has a very lively mind and you should read anything he writes.

195 **Dietmar Kuhl and Paul Worley:** They headed two labs that independently made this discovery. See Wolfgang Link et al., "Somatodendritic Expression of an Immediate Early Gene is Regulated by Synaptic Activity," *PNAS* 92, no. 12 (June 1995): 5734–38, doi:10.1073/pnas.92.12.5734; and Gregory L. Lyford et al., "*Arc*, a Growth Factor and Activity-Regulated Gene, Encodes a Novel Cytoskeleton-Associated Protein That Is Enriched in Neuronal Dendrites," *Neuron* 14, no. 2 (February 1995): 433–45, doi:10.1016/0896-6273(95)90299-6.

196 **invades other nearby neurons:** This has been shown in mice and flies; see, for example, a summary and references therein: Sara Reardon, "Cells Hack Virus-Like Protein to Communicate," *Nature* News, January 11, 2018, doi:10.1038/d41586-018-00492-w.

199 **proposed *constructor theory*:** Several further works have since been published, but a starting point is: David Deutsch, "Constructor Theory," *Synthese* 190 (April 2013): 4331–59, doi:10.1007/s11229-013-0279-z.

## 7. LIFE MADE MACHINE

201 **Hod Lipson's laboratory:** If you want an easy introduction to some of Hod's earlier work then I'd suggest watching his TED Talk from 2007: www.ted.com/talks/hod_lipson_building_self_aware_robots.

202 ***the Economic World*:** Kevin Kelly's book is available online as well as in print: https://kk.org/outofcontrol/.

203 **the study of machine behavior:** This popular-level introduction is useful: John Pavlus, "The Anthropologist of Artificial Intelligence," *Quanta Magazine*, August 26, 2019, www.quantamagazine.org/iyad-rahwan-is-the-anthropologist-of-artificial-intelligence-20190826; and the proposal for a new field of study: Iyad Rahwan et al., "Machine Behaviour," *Nature* 568 (April 2019): 477–86, doi:10.1038/s41586-019-1138-y.

203 **The RAND Corporation thinks:** See, for example, its site with a number of studies: www.rand.org/topics/autonomous-vehicles.html.

203 **self-replicating AI:** See, for example: Oscar Chang and Hod Lipson, "Neural Network Quine," *ALIFE 2018: Proceedings of the Artificial Life Conference 2018* 30, MIT Press Journal (July 2018), 234–41, doi:10.1162/isal_a_00049.

204 **his "cellular robot.":** See, for example: Toshio Fukuda et al., "Concept of Cellular Robotic System (CEBOT) and Basic Strategies for Its Realization," *Computers & Electrical Engineering* 18, no. 1 (January 1992): 11–39, doi:10.1016/0045-7906(92)90029-D.

207 **John von Neumann and Alan Turing:** So much has been written and said about these two that it's hard to point to any particular set of source material. A few minutes interrogating the dataome online will lead you to many troves of articles and snippets.

207 **computer called EDVAC:** This report is available in a number of places; see, for example, a scanned version at the Smithsonian Libraries: https://library.si.edu/digital-library/book/firstdraftofrepo00vonn. John von Neumann, "First Draft of a Report on the EDVAC," Contract no. N-670-ORD-4926 between the United States Army Ordnance Department and the University of Pennsylvania (Philadelphia: Moore School of Electrical Engineering, University of Pennsylvania, 1945), doi:10.5479/sil.538961.39088011475779.

208 **von Neumann architecture:** More specifically, the memory sits alongside an arithmetic logic unit (the heart of a CPU), a control unit (lining up processes in a register, keeping count of programs and cycles), and an input/output unit (a way to shuffle data in and out, perhaps simultaneously). Most of these ideas were, in essence, already there in Charles Babbage's earlier mechanical computers.

208 **that he published in 1946:** Specifically, his paper on the Automatic Computing Engine (ACE)—in which he also acknowledges Herman Hollerith's punched card systems. See, for example, a nice reprinting of this and companion papers: Alan Turing and Michael Woodger, *A. M. Turing's ACE Report of 1946 and Other Papers*, (Cambridge, MA: MIT Press, 1986); or the materials online at: www.alanturing.net/turing_archive/archive/index/aceindex.html.

208 **Already in 1936:** In many ways the most famous of Turing's earlier works, and way ahead of its time: A. M. Turing, "On Computable Numbers, with an Application to the Entscheidungsproblem," *Proceedings of the London Mathematical Society* s2-42, no. 1 (January 1937): 230–65, doi:10.1112/plms/s2-42.1.230.

209 **computer system from unreliable components:** John von Neumann, "Lectures on Probabilistic Logics and the Synthesis of Reliable Organisms from Unreliable Components," Lecture series at Caltech, 1952, online at: https://static.ias.edu/pitp/archive/2012files/Probabilistic_Logics.pdf.

209 **zebra stripes to the shape of bodies:** This is much less well known, but another piece of brilliance from Turing: Alan Mathison Turing, "The Chemical Basis of Morphogenesis," *Philosophical Transactions of the Royal Society B* 237, no. 641 (August 1952): 37–72, doi:10.1098/rstb.1952.0012.

209 **known as the Turing test:** The paper is: A. M. Turing, "Computing Machinery and Intelligence," *Mind* 59, no. 236 (October 1950): 433–60, doi:10.1093/mind/LIX.236.433.

211 **well known in recent years:** It has even had the Hollywood treatment in the excellent (albeit limited in depth) movie *The Imitation Game* in 2014.

217 **a planet of machines and data:** I debated long and hard whether or not to commit this "thought experiment" to the page. It can seem silly, but I am trying to get the reader to genuinely reframe their sense of the world, down to an emotional level, and a story seems like a perfectly reasonable way to do this.

220 **wannabe alchemist Hennig Brand:** The story of his work on phosphorus is quite complex, and involves some of the big names of the time, like Leibniz. A terrific essay on all of this was written by the chemist and historian of science Mary Weeks back in 1933: Mary Elvira Weeks, "The Discovery of the Elements. XXI. Supplementary Note on the Discovery of Phosphorus," *Journal of Chemical Education* 10, no. 5 (May 1933): 302, doi:10.1021/ed010p302.

221 **one such effort:** For another example see: Stephane Doncieux et al., "Evolutionary Robotics: What, Why, and Where To," *Frontiers in Robotics and AI* 2 (March 2015): 4, doi:10.3389/frobt.2015.00004.

221 **a group led by Robert Grass:** The scientific report on this is: Julian Koch et al., "A DNA-of-Things Storage Architecture to Create Materials with Embedded Memory," *Nature Biotechnology* 38 (January 2020): 39–43, doi:10.1038/s41587-019-0356-z.

225 **23 billion domestic chickens:** In fact it has been proposed that chicken remains will be one of the primary pieces of fossil evidence in the future for the presence of humans and our impact on Earth's environment: Carys E. Bennett et al., "The

Broiler Chicken as a Signal of a Human Reconfigured Biosphere," *Royal Society Open Science* 5, no. 12 (December 2018): 180325, doi:10.1098/rsos.180325.

227 **experiment reported in 2011:** See Levi T. Morran et al., "Running with the Red Queen: Host-Parasite Coevolution Selects for Biparental Sex," *Science* 333, no. 6039 (July 2011): 216–18, doi:10.1126/science.1206360.

## 8. THE GREAT BLENDING

231 **Niels Bohr, speaking:** Said to Wolfgang Pauli after his presentation of Heisenberg's and Pauli's nonlinear field theory of elementary particles, at Columbia University (1958), as reported by Freeman J. Dyson in his paper "Innovation in Physics," *Scientific American* 199, No. 3 (September 1958): 74–82; reprinted in *JingShin Theoretical Physics Symposium in Honor of Professor Ta-You Wu*, ed. Jong-Ping Hsu and Leonardo Hsu, 73–90 (Singapore and River Edge, NJ: World Scientific, 1998), 84.

233 **Siderian through the Cambrian:** The Siderian starts at 2.5 billion years ago and runs until 2.3 billion years ago (there are nine other periods until the Cambrian at 541 million years until 485.4 million years ago). Prior to 2.5 billion years ago we simply haven't named any periods, but we have eons: the Archaean eon from 4 billion years ago until 2.5 billion years, and the earlier Hadean eon, from the formation of Earth around 4.6 billion years ago until 4 billion years ago.

234 **overlapping branches of life:** See for example: Rachel Wood et al., "Integrated Records of Environmental Change and Evolution Challenge the Cambrian Explosion," *Nature Ecology and Evolution* 3 (March 2019): 528–38, doi:10.1038/s41559-019-0821-6. But there is also evidence for a "short" 20-million-year span for the "explosion"; see, for example: John R. Paterson, Gregory D. Edgecombe, and Michael S. Y. Lee, "Trilobite Evolutionary Rates Constrain the Duration of the Cambrian Explosion," *PNAS* 116, no. 10 (March 2019): 4394–99, doi:10.1073/pnas.1819366116.

234 **the griffinfly genus:** These were truly formidable insects. Recent research suggests they likely inhabited relatively open terrain and were "hawkers," with attributes enabling them to catch prey in the air: André Nel et al., "Palaeozoic Giant Dragonflies Were Hawker Predators," *Scientific Reports* 8 (August 2018): 12141, doi:10.1038/s41598-018-30629-w.

235 **known as the Great Dying:** It is such a feature in the geologic and paleontological record that it's been discussed, dissected, and re-discussed and re-dissected many times, and still is. There are also numerous popular-level accounts—see, for example: Michael J. Benton, *When Life Nearly Died: The Greatest Mass Extinction of All Time* (New York: Thames & Hudson, 2003).

236 **sous-vide bisque:** Evidence for dramatic global temperature rise: Yadong Sun et al., "Lethally Hot Temperatures During the Early Triassic Greenhouse," *Science* 338, no. 6105 (October 2012): 366–70, doi:10.1126/science.1224126.

236 **to occupy and innovate in:** See, for example: Sarda Sahney and Michael J. Benton, "Recovery from the Most Profound Mass Extinction of All Time," *Proceedings of the Royal Society B* 275, no. 1636 (April 2008): 759–65, doi:10.1098/rspb.2007.1370.

237 **free-swimming species:** See, for example: Haijun Song, Paul B. Wignall, and Alexander M. Dunhill, "Decoupled Taxonomic and Ecological Recoveries from the Permo-Triassic Extinction," *Science Advances* 4, no. 10 (October 2018): eaat5091, doi:10.1126/sciadv.aat5091.

237 **"The Energy Expansions of Evolution.":** Olivia P. Judson, "The Energy Expansions of Evolution," *Nature Ecology and Evolution* 1 (April 2017): 0138, doi:10.1038/s41559-017-0138.

238 **reeking with the stuff:** There is a very large scientific literature on the history of oxygen on Earth and its many roles. A nice overview is Heinrich D. Holland, "The Oxygenation of the Atmosphere and Oceans," *Philosophical Transactions of the Royal Society B* 361, no. 1470 (June 2006): 903–15, doi:10.1098/rstb.2006.1838.

238 **only possible exceptions:** A few of these have come to light, including a salmon parasite that doesn't have mitochondria, and some other tiny marine organisms in the dark oceanic depths. See, for example: Marek Mentel and WIlliam Martin, "Anaerobic Animals from an Ancient, Anoxic Ecological Niche," *BMC Biology* 8 (April 2010): 32, doi:10.1186/1741-7007-8-32.

240 **A study in 2019:** See Lewis J. Alcott, Benjamin J. W. Mills, and Simon W. Poulton, "Stepwise Earth Oxygenation is an Inherent Property of Global Biogeochemical Cycling," *Science* 366, no. 6471 (December 2019): 1333–37, doi:10.1126/science.aax6459.

242 **compound called NADH:** A chemically reduced form of nicotinamide adenine dinucleotide is a funky-looking molecule consisting of two nucleotides holding hands.

242 **Lynn Margulis in 1967:** At the time she was married to Carl Sagan and used that surname; see: Lynn Sagan, "On the Origin of Mitosing Cells," *Journal of Theoretical Biology* 14, no. 3 (March 1967): 255–74, doi:10.1016/0022-5193(67)90079-3.

243 **hit the jackpot:** This way of looking at things—flipping the normal assumptions and order of cause-and-effect—takes some getting used to, but I think it's very helpful even if, in the end, it is only a way to shake our thinking loose. Some of the ideas here came together because of a conversation I had with Bryan Johnson, the entrepreneur and CEO of Kernel (a pioneer in producing neural interfaces). Bryan uses the term "anti-familiar" and it's a great description of these thought inversions.

243 **plenty of internal structures:** See for example: Megan J. Dobro et al. "Uncharacterized Bacterial Structures Revealed by Electron Cryotomography," *Journal of Bacteriology* 199, no. 17 (August 2017): e00100-17, doi10.1128/JB.00100-17. And a popular-level article: Jordana Cepelewicz, "Bacterial Complexity Revises Ideas About 'Which Came First?,'" *Quanta Magazine,* June 12, 2019, www.quantamagazine.org/bacterial-organelles-revise-ideas-about-which-came-first-20190612/.

246 **catching Isaac Newton's attention:** It didn't really, or at least it seems that Newton invented this story as he tried to solidify his place in history. His first biographer,

William Stukeley, recounts Newton bringing this up over tea on a hot day in 1726: "... he told me, he was just in the same situation, as when formerly, the notion of gravitation came into his mind. Why sh[oul]d that apple always descend perpendicularly to the ground, thought he to himself; occasion'd by the fall of an apple, as he sat in a contemplative mood. . . ." See the memoir written by Stukeley online, digitized at: https://royalsociety.org/collections/turning-pages/.

249 **and Charles Cockell in 2015:** See Hanna K. E. Landenmark, Duncan H. Forgan, and Charles S. Cockell, "An Estimate of the Total DNA in the Biosphere," *PLoS Biology* 13, no. 6 (June 2015): e1002168, doi:10.1371/journal.pbio.1002168.

249 **$10^{21}$ to $10^{22}$ FLOPS:** Numbers like these are necessarily estimates. Here I've taken some from a few years ago and extrapolated a little, but I think this is "ballpark" accurate. See for example: https://aiimpacts.org/global-computing-capacity/ (accessed January 2020).

## 9. A UNIVERSE OF DATAOMES

255 **Nobel Prize in Literature acceptance speech:** Morrison was a special human being; the text of her speech can be found at the Nobel Prize website: www.nobelprize.org/prizes/literature/1993/morrison/lecture/. Interestingly, in this speech she uses a tale of a blind person asked to identify whether a bird is living or dead. She also gives a brilliant analysis of the nature of language itself, as a thing that can govern us as much as we use it to govern the world. You should go read this and do your part for the dataome.

260 **chemical pressure or potential:** See for example this essay: Harold Morowitz and Eric Smith, "Energy Flow and the Organization of Life," *Complexity* 13, no. 1 (September/October 2007): 51–59, doi:10.1002/cplx.20191.

260 **patterned-yet-unpredictable way:** These pattern-making blobs are part of Rayleigh–Bénard convection, or instability, and a classic example of self-organizing nonlinear systems.

260 **non-equilibrium processes:** Indeed, this is not a frivolous statement, it is quantifiably true: Stephen Ornes, "Core Concept: How Nonequilibrium Thermodynamics Speaks to the Mystery of Life," *PNAS* 114, no. 3 (January 2017): 423–24, doi:10.1073/pnas.1620001114.

261 **a thermodynamic imperative:** One nice thing about Jeremy England's ideas is that they can be tested (and thus far seem to hold up, at least in simulations). Two sources on this are: Jordan M. Horowitz and Jeremy L. England, "Spontaneous Fine-Tuning to Environment in Many-Species Chemical Reaction Networks," *PNAS* 114, no. 29 (July 2017): 7565–70, doi:10.1073/pnas.1700617114; and Tal Kachman, Jeremy A. Owen, and Jeremy L. England, "Self-Organized Resonance during Search of a Diverse Chemical Space," *Physical Review Letters* 119, no. 3 (July 2017): 038001, doi:10.1103/PhysRevLett.119.038001.

262 **beautiful study published in 2018:** The paper is: Marius Somveille, Ana S. L. Rodrigues, and Andrea Manica, "Energy Efficiency Drives the Global Seasonal Distribution of Birds," *Nature Ecology and Evolution* 2 (May 2018): 962–69, doi:10.1038/s41559-018-0556-9, and it's really lovely. The authors have also done further work on this, including looking at the past 50,000 years of potential migratory routes.

263 **in his talk title:** Feynman's lecture has become almost a cliché in nanoscience (with the occasional call to not reference it at the start of every new research article), but it remains important and insightful. You can find it online in many places: Richard P. Feynman, "There's Plenty of Room at the Bottom: An Invitation to Enter a New Field of Physics," *Engineering and Science* 23, no. 5 (February 1960): 22–36. For example https://web.pa.msu.edu/people/yang/RFeynman_plentySpace.pdf.

264 **the potential isn't extraordinary:** A terrific overview is given by George Musser, "Job One for Quantum Computers: Boost Artificial Intelligence," *Quanta Magazine,* January 29, 2018, www.quantamagazine.org/job-one-for-quantum-computers-boost-artificial-intelligence-20180129/.

265 **very short distances:** It is still a matter of research and debate. See, for example: Erling Thyrhaug et al, "Identification and Characterization of Diverse Coherences in the Fenna–Matthews–Olson Complex," *Nature Chemistry* 10 (May 2018): 780–86, doi:10.1038/s41557-018-0060-5.

267 **wrote about it in a 2003 paper:** Nick Bostrom, "Are We Living in a Computer Simulation?," *The Philosophical Quarterly* 53, no. 211 (April 2003): 243–55, doi:10.1111/1467-9213.00309.

268 **MIT physicist Seth Lloyd:** See, for example, his popular-level book: Seth Lloyd, *Programming the Universe: A Quantum Computer Scientist Takes on the Cosmos* (New York: Alfred A. Knopf, 2006).

268 **planet-discovering bender:** Currently the number of likely candidate detections of planets around other stars sits well above 4,000—although the amount of information we have about any of these planets is extremely limited, barely telling us their orbital configurations and sizes. That will change in the near future.

272 **assert itself like any other:** I actually find this to be a fascinating question: If an entity like a virus "infects" our dataome through our efforts to sequence its genes and thwart its pathogenic effects on us, has it gone through its own viral form of transcendence? Arguably it has. Perhaps there's a clue in here to the nature of life as that propagating river of information.

273 **Freeman Dyson in 1960:** Those were the days, when a single page of wild projection and extrapolation could be published as a scientific article. Of course it helped that Dyson's scientific credentials were impeccable: Freeman J. Dyson, "Search for Artificial Stellar Sources of Infrared Radiation," *Science* 131, no. 3414 (June 1960): 1667–68, doi:10.1126/science.131.3414.1667.

273 **physicist Fred Adams:** Adams talked about this at a scientific meeting organized by FQXi, and there's a podcast: https://fqxi.org/community/forum/topic/3317.

275 **probes back and forth:** Can this really work? Does the time required to send queries or to retrieve information (by probe, for example) actually undo any gain from the different clock speeds? I believe it is traditional to leave these kinds of questions as an exercise for the reader.

278 **a "Boltzmann Brain.":** This idea gets bandied about quite a bit, and quite often misused—as people forget that it is actually kind of meant to be absurd, that is its purpose. The physicist Sean Carroll has written some nice stuff about this. See, for example: Sean M. Carroll, "Why Boltzmann Brains Are Bad," arXiv.org, February 2017, https://arxiv.org/abs/1702.00850.

279 **in his 1936 work:** Albert Einstein, "Physics and Reality," *Journal of the Franklin Institute* 221, no. 3 (March 1936): 349–82, doi:10.1016/S0016-0032(36)91047-5.

280 **near-seventy-page-long essay:** This essay is quite a ride: Scott Aaronson, "The Ghost in the Quantum Turing Machine," in *The Once and Future Turing: Computing the World*, ed. S. Barry Cooper & Andrew Hodges, 193–296 (Cambridge, UK: Cambridge University Press, 2016).

280 **after one Frank Knight:** He was an American economist (1885–1972), and his idea of "true uncertainty" is still widely referenced and discussed.

280 **Nassim Taleb:** See his book: Nassim Nicholas Taleb, *The Black Swan: The Impact of the Highly Improbable* (New York: Random House, 2007)..

283 **any kind of bits:** The physicist Seth Lloyd also did this: Seth Lloyd, "Computational Capacity of the Universe," *Physical Review Letters* 88, no. 23 (June 2002): 237901, doi:10.1103/PhysRevLett.88.237901.

288 **an ongoing participant:** Indeed, we are all inoculated with information and induced into being carriers and maintainers of the dataome from a very early stage, beginning at birth.

289 **books, *Who Owns the Future?*:** The book is: Jaron Lanier, *Who Owns the Future?* (New York: Simon & Schuster, 2013).

290 **the Domesday Book:** There is a marvelous resource on this at the UK's National Archives: www.nationalarchives.gov.uk/domesday/discover-domesday/.

291 **There is a tide:** I want to acknowledge that this quote from Shakespeare came up in conversation with Dmitri Gunn, who runs the incredibly successful TEDxCambridge program and who has had many a patient conversation with me as I felt my way through the material of this book.

# Index